Peter E. Nielsen (Ed.)

Pseudo-peptides in Drug Discovery

Further Titles of Interest

Chi-Huey Wong (Ed.)

Carbohydrate-based Drug Discovery

2003. ISBN 3-527-30632-3

M. Demeunynck, C. Bailly, W. D. Wilson (Eds.)

DNA and RNA Binders

2002. ISBN 3-527-30595-5

N. Sewald, H.-D. Jakubke

Peptides: Chemistry and Biology

2002. ISBN 3-527-30405-3

K. C. Nicolaou, R. Hanko, W. Hartwig (Eds.)

Handbook of Combinatorial Chemistry

2002. ISBN 3-527-30509-2

Peter E. Nielsen (Ed.)

Pseudo-peptides in Drug Discovery

WILEY-VCH

WILEY-VCH Verlag GmbH & Co. KGaA

Prof. Dr. Peter E. Nielsen
Department of Medical Biochemistry and Genetics
University of Copenhagen
The Panum Institute
Blegdamsvej 3c
2200 Copenhagen N
Denmark

Library of Congress Card No.: applied for

British Library Cataloguing-in-Publication Data
A catalogue record for this book is available from the British Library.

Bibliographic information published by Die Deutsche Bibliothek
Die Deutsche Bibliothek lists this publication in the Deutsche Nationalbibliografie; detailed bibliographic data is available in the Internet at <http://dnb.ddb.de>

© 2004 WILEY-VCH Verlag GmbH & Co. KGaA, Weinheim, Germany

Printed in the Federal Republic of Germany
Printed on acid-free paper

Cover Grafik-Design Schulz, Fußgönheim
Composition K+V Fotosatz GmbH, Beerfelden
Printing Strauss Offsetdruck GmbH, Mörlenbach
Bookbinding Litges & Dopf Buchbinderei GmbH, Heppenheim

ISBN 3-527-30633-1

Contents

Preface

Peptides are used extensively by Nature for a variety of signalling functions in both unicellular and multicellular organisms including man. For example, many peptide hormones and analogous short peptides exert their action by binding to membrane receptors. Peptides may also show biological activities in the form of nerve toxins, antibacterial agents or general cell toxins. Therefore, it is no wonder that medical drug discovery has extensively exploited peptides as lead compounds. This development was further accelerated by the development of very effective methods for solid phase synthesis of peptides, and in particular the development of combinatorial methods for synthesizing and screening peptide libraries.

However, most natural peptides are composed of L-form a-amino acids and because of the ubiquitous prevalence of peptidases they have limited biostability, and consequently low bioavailability. Thus, a novel field of peptidomimetics has emerged in drug discovery, in attempts to design non-peptide compounds mimicking the pharmacophore and thus the activity of the original peptide.

This field has also inspired the development of a range of pseudo-peptides, that is polyamides composed of amino acids other than a-amino acids. These include for instance peptoids, β-amino acid oligomers and also compounds such as peptide nucleic acids and DNA binding polyamides, all of which share the amide (peptide) chemistry with natural peptides.

The present book attempts to present the state of the art in the rapidly expanding field of pseudo-peptides, in particular relating to (long term aims of) drug discovery. Many chemists are realizing the power and versatility of "peptide" chemistry, and the large structural and functional space attainable using this technology. Hopefully the book will inspire new developments.

I am extremely grateful to the friends and colleagues who have made this project possible by investing their time and expertise.

Copenhagen, October 2003   *Peter E. Nielsen*

List of Contributors

Annelise E. Barron
Department of Chemical Engineering
Northwestern University
2145 Sheridan Road
Tech E136
Evanston, IL 60208
USA
a-barron@northwestern.edu

Frederik Beck
GEA A/S
Kanalholmen 8–12
DK 2650 Hvidovre
Denmark

Peter B. Dervan
Division of Chemistry and Chemical
Engineering
California Institute of Technology
Pasadena, CA 91125
USA
dervan@caltech.edu

Subhakar Dey
Department of Chemistry
Case Western Reserve University
10900 Euclid Avenue
Cleveland, OH 44106
USA

Benjamin S. Edelson
Division of Chemistry and Chemical
Engineering
California Institute of Technology
Pasadena, CA 91125
USA

Eric J. Fechter
Division of Chemistry and Chemical
Engineering
California Institute of Technology
Pasadena, CA 91125
USA

Philip P. Garner
Department of Chemistry
Case Western Reserve University
10900 Euclid Avenue
Cleveland, OH 44106
USA
ppg@cwru.edu

Joel M. Gottesfeld
Department of Molecular Biology
The Scripps Research Institute
10550 North Torrey Pines Road
La Jolla, CA 92037
USA
joelg@scripps.edu

GILLES GUICHARD
Institut de Biologie Moléculaire
et Cellulaire
UPR 9021 CNRS
Université Louis Pasteur Strasbourg
15, rue Descartes
67084 Strasbourg Cedex
France
g.guichard@ibmc.u-strasbg.fr

YUMEI HUANG
Department of Chemistry
Case Western Reserve University
10900 Euclid Avenue
Cleveland, OH 44106
USA

KENT KIRSHENBAUM
Department of Chemistry
New York University
100 Washington Square East
Room 1001
New York, NY 10003
USA

UFFE KOPPELHUS
Department of Medical Biochemistry
and Genetics
University of Copenhagen
The Panum Institute
Blegdamsvej 3c
DK 2200 Copenhagen N
Denmark

PETER E. NIELSEN
University of Copenhagen
The Panum Institute
Blegdamsvej 3c
DK 2200 Copenhagen N
Denmark
pen@imbg.ku.dk

JAMES A. PATCH
Department of Chemical Engineering
Northwestern University
2145 Sheridan Road
Tech E136
Evanston, IL 60208
USA

ALESSANDRO SCARSO
Department of Organic Chemistry
and ITM-CNR Padova Section
University of Padova
Via Marzolo 1
35131 Padova
Italy

PAOLO SCRIMIN
Department of Organic Chemistry and
ITM-CNR Padova Section
University of Padova
Via Marzolo 1
35131 Padova
Italy
paolo.scrimin@unipd.it

SHANNON L. SEURYNCK
Department of Chemical Engineering
Northwestern University
2145 Sheridan Road
Tech E136
Evanston, IL 60208
USA

RONALD N. ZUCKERMANN
Bioorganic Chemistry
Chiron Corp.
4560 Horton St.
Emeryville, CA 94608
USA

1
Versatile Oligo(*N*-Substituted) Glycines:
The Many Roles of Peptoids in Drug Discovery

James A. Patch, Kent Kirshenbaum, Shannon L. Seurynck, Ronald N. Zuckermann, and Annelise E. Barron

1.1
Introduction

Despite their wide range of important bioactivities, polypeptides are generally poor drugs. Typically, they are rapidly degraded by proteases *in vivo*, and are frequently immunogenic. This fact has stimulated widespread efforts to develop peptide mimics for biomedical applications, a task that presents formidable challenges in molecular design. Chemists seek efficient routes to peptidomimetic compounds with enhanced pharmacological properties, which retain the activities of their biological counterparts. Since peptides play myriad roles in living systems, it is likely that no individual strategy will suffice. Indeed, a wide variety of different peptidomimetic oligomer scaffolds have been explored [1]. In order to address multiple design criteria for applications ranging from medicinal chemistry to materials science, researchers have worked to identify a non-natural chemical scaffold that recapitulates the desirable attributes of polypeptides. These include good solubility in aqueous solution, access to facile sequence-specific assembly of monomers containing chemically diverse side chains, and the capacity to form stable, biomimetic folded structures.

Among the first reports of chemically diverse peptide mimics were those of (*N*-substituted) glycine oligomers (peptoids) [2]. Sequence-specific oligopeptoids have now been studied for over a decade, and have provided illustrative examples of both the potential of peptidomimetics and the obstacles faced in translating this potential into clinically useful compounds. We begin this chapter with a summary of the desirable attributes of peptoids as peptide mimics, along with a description of strategies for their chemical synthesis. Throughout the remainder of the chapter, we present an overview of biomedically relevant studies of peptoids with an emphasis on recently reported results. The chapter includes discussion of peptoid combinatorial libraries, folded peptoid structures, and biomimetic peptoid sequences. Finally, we conclude by suggesting promising avenues for future investigations.

Peptoids are an archetypal and relatively conservative example of a peptidomimetic oligomer (Tab. 1.1). In fact, the sequence of atoms along the peptoid backbone is identical to that of peptides. However, peptoids differ from peptides in the manner of side chain appendage. Specifically, the side chains of peptoid oligo-

Pseudo-peptides in Drug Discovery. Edited by Peter E. Nielsen
Copyright © 2004 Wiley-VCH Verlag GmbH & Co. KGaA, Weinheim
ISBN: 3-527-30633-1

Tab. 1.1 Comparison of key characteristics of peptides and peptoids

	Peptides	*Peptoids*
Identity	Sequence-specific polymers of amino acids	Sequence-specific polymers of N-substituted glycines
Synthesis	Solid-phase. Polymerization of N-protected α-amino acids (Fmoc or Boc).	Solid-phase sub-monomer synthesis
Number of side chains?	20+	Derived from hundreds of available primary amines
Secondary structures	Helices (3^{10}, a), β-sheets	Helices
Structural stabilization	Intra-chain hydrogen bonds	Steric and electronic repulsions
Thermal stability of structures?	Up to $\sim 40\,^{\circ}$C	$>75\,^{\circ}$C
Structural stability to solvent environment?	May denature in high salt, organic solvent, or at pH extremes	Generally stable to salt, pH, and organic solvent
In vivo stability	Rapidly degraded (proteolysis)	Stable to proteolysis, excreted whole in urine

mers are shifted to become pendant groups of the main-chain nitrogen atoms (Fig. 1.1). The presentation of peptide and peptoid side chains is roughly isosteric, potentially allowing for suitable mimicry of the spacing between the critical chemical functionalities of bioactive peptides. Peptoid monomers are linked through polyimide bonds, in contrast to the amide bonds of peptides (with the sole exception of proline residues, which are also -imino acids). Peptoids lack the hydrogen of the peptide secondary amide, and are thus incapable of forming the same types of hydrogen bond networks that stabilize peptide helices and β-sheets, respectively. The peptoid oligomer backbone is achiral; however, chiral centers can be included in the side chains to obtain secondary structures with a preferred handedness [3, 4]. In addition, peptoids carrying N-substituted versions of the proteinogenic side chains are highly resistant to degradation by proteases [5], which is an important attribute of a pharmacologically useful peptide mimic.

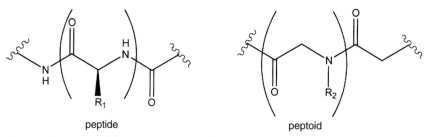

peptide   peptoid

Fig. 1.1 Comparison of the primary structure of peptide and peptoid oligomers

The efficient solid-phase synthesis of peptoids (Section 1.2) enables facile combinatorial library generation. In the "sub-monomer" synthetic strategy, peptoid monomers are synthesized by a two-step process from a haloacetic acid and a primary amine. A wide variety of amines are commercially available, which facilitates the incorporation of chemically diverse side chains. High synthetic yields typically attained at each synthetic step permit the propagation of peptoid chains to substantial lengths. For instance, peptoids up to 48 residues long have been synthesized with reasonable yields of the full-length target sequence [6].

1.2
Peptoid Synthesis

1.2.1
Solid-Phase Synthesis

Sequence-specific heteropolymers, as a class of synthetic molecules, are unique in that they must be made by chemical steps that add one monomer unit at a time. Moreover, to create truly protein-like structures, which typically have chain lengths of at least 100 monomers and a diverse set of 20 side chains (or more), extremely efficient and rapid couplings under general reaction conditions are necessary. For these reasons, solid-phase synthesis is typically used, so that excess reagents can be used to drive reactions to completion, and subsequent reaction work-ups are quite rapid.

A common feature of most solid-phase oligomer syntheses (e.g. peptide, oligonucleotide, peptide nucleic acid, β-peptide, etc.) is that they are made by a two-step monomer addition cycle. First, a protected monomer unit is coupled to a terminus of the resin-bound growing chain, and then the protecting group is removed to regenerate the active terminus. Each side chain group requires a separate N^α-protected monomer. The first oligopeptoids reported were synthesized by this method, for which a set of Fmoc-protected peptoid monomers was made [2].

Specifically, the carboxylates of N^α-Fmoc-protected (and side chain-protected) N-substituted glycines were activated and then coupled to the secondary amino

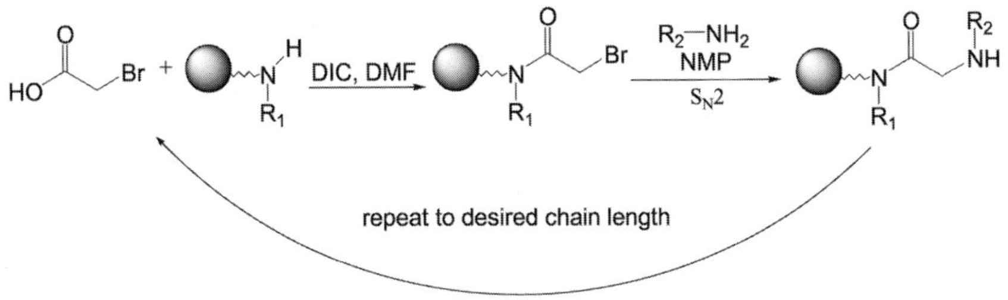

Fig. 1.2 Solid-phase sub-monomer peptoid synthesis

group of a resin-bound peptoid chain. Removal of the Fmoc group was then followed by addition of the next monomer. Thus, peptoid oligomers can be thought of as condensation homopolymers of *N*-substituted glycine. There are several advantages to this method [7], but the extensive synthetic effort required to prepare a suitable set of chemically diverse monomers is a significant disadvantage of this approach. Additionally, the secondary *N*-terminal amine in peptoid oligomers is more sterically hindered than the primary amine of an amino acid, which slows coupling reactions.

1.2.2
Sub-monomer Solid-Phase Method

A major breakthrough came in 1992 when a much more efficient method of peptoid synthesis was invented [8]. In this method, each *N*-substituted glycine (NSG) monomer is assembled from two readily available "sub-monomers" in the course of extending the NSG oligomer [9]. This method is known as the sub-monomer method, in which each cycle of monomer addition consists of two steps, an acylation step and a nucleophilic displacement step (Fig. 1.2). Thus, peptoid oligomers can also be considered to be alternating condensation copolymers of a haloacetic acid and a primary amine. This method is unique among solid-phase oligomer syntheses in that there are no protecting groups used in elongating the main chain. As in the original method, the direction of oligomer synthesis utilizing these sub-monomers occurs in the carboxy to amino direction.

In the first step, a resin-bound secondary amine is acylated with bromoacetic acid, in the presence of *N,N*-diisopropylcarbodiimide. Acylation of secondary amines is difficult, especially when coupling an amino acid with a bulky side chain. The sub-monomer method, on the other hand, is facilitated by the use of bromoacetic acid, which is a very reactive acylating agent. Activated bromoacetic acid is bis-reactive, in that it acylates by reacting with a nucleophile at the carbonyl carbon, or it can alkylate by reacting with a nucleophile at the neighboring aliphatic carbon. Because acylation is approximately 1000 times faster than alkylation, acylation is exclusively observed.

The second step introduces the side chain group by nucleophilic displacement of the bromide (as a resin-bound *a*-bromoacetamide) with an excess of primary amine. Because there is such diversity in reactivity among candidate amine sub-monomers, high concentrations of the amine are typically used (~ 1–2 M) in a polar aprotic solvent (e.g. DMSO, NMP or DMF). This S_N2 reaction is really a mono-alkylation of a primary amine, a reaction that is typically complicated by over-alkylation when amines are alkylated with halides in solution. However, since the reactive bromoacetamide is immobilized to the solid support, any over-alkylation side-products would be the result of a cross-reaction with another immobilized oligomer (slow) in preference to reaction with an amine in solution at high concentration (fast). Thus, in the sub-monomer method, the solid phase serves not only to enable a rapid reaction work-up, but also to isolate reactive sites from one another. However, the primary advantage of the method is that each primary amine is much simpler in structure than a protected full monomer. The fact that several hundred primary amines are commercially available greatly facilitates peptoid synthesis.

Most primary amines that are neither sterically hindered nor very weak nucleophiles will incorporate into the peptoid chain in high yield. However, protection of reactive side-chain functionalities such as carboxyl, thiol, amino, hydroxy and other groups may be required to minimize undesired side reactions [10]. Acid-labile protecting groups are preferred, as they may be removed during peptoid cleavage from solid support with trifluoroacetic acid (TFA). The mild reactivity of some side-chain moieties toward displacement or acylation allows their use without protection in some cases (e.g. indole, phenol). In these cases, the side chain may become transiently acylated during the acylation step and will subsequently revert back to the free side chain upon treatment with the amine in the displacement step. Heterocyclic side-chain moieties such as imidazole, pyrazine, quinolines and pyridines may also become transiently acylated. However, since these groups are more nucleophilic they are susceptible to alkyation by the activated bromoacetic acid. In these cases, clean incorporation can be achieved by replacing bromoacetic acid with chloroacetic acid [11].

1.2.3
Side Reactions

There are a few competing side reactions that are unique to the synthesis of peptoid oligomers. For example, peptoid dimer synthesis often leads to formation of the cyclic diketopiperazines instead of the linear molecule [12]. Sub-monomers whose side chains bear a nucleophile three or four atoms from the amino nitrogen, are also prone to cyclizations after bromoacetylation. We have found the sub-monomers trityl-histamine, 2-(aminomethyl)benzimidazole, and 2-aminoethylmorpholine fall into this category. Electron-rich benzylic side chains, such as those derived from p-methoxy-1-phenylethylamine or 2,4,6-trimethoxybenzylamine, will fall off the main chain during the post-synthetic acidolytic cleavage with TFA.

1.2.4
Post-Synthetic Analysis

Like peptide oligomers, peptoids can be analyzed by HPLC and by mass spectrometry. They can be sequenced by Edman degradation [13] or by tandem mass spectrometry [14] since, like polypeptides, they conveniently fragment along the main chain amides [15, 16].

1.3
Drug Discovery via Small-Molecule Peptoid Libraries

Because of their ease of synthesis and their structural similarity to peptides, many laboratories have used peptoids as the basis for combinatorial drug discovery. Peptoids were among the first non-natural compounds used to establish the basic principles and practical methods of combinatorial discovery [17]. Typically, diverse libraries of relatively short peptoids (<10 residues) are synthesized by the mix-and-split method and then screened for biological activity. Individual active compounds can then be identified by iterative re-synthesis, sequencing of compounds on individual beads, or indirect deduction by the preparation of positional scanning libraries.

1.3.1
Peptoid Drugs from Combinatorial Libraries

Early on, peptoid trimers that were nanomolar ligands for the opiate and a_1-adrenergic receptors were identified by *in vitro* receptor binding experiments [17] (Fig. 1.3). These ligands were discovered from a library of 3500 trimers by iterative re-synthesis, where activity was followed through successive rounds of re-synthesis of smaller pools. These compounds showed some *in vivo* activity, despite their poor pharmacokinetic properties [18]. Trimeric peptoid libraries were also screened for antimicrobial activity in whole cell assays, which yielded compounds with modest activities against *S. aureus* and *E. coli* [19]. Very large libraries of peptoid trimers ($\sim 350,000$ compounds) were screened against a variety of agrochemical targets to provide a sub-micromolar antagonist of the nicotinic acetylcholine receptor, following iterative deconvolution [20]. Positional scanning libraries of peptoid trimers were used to discover noncompetitive antagonists of the vanilloid receptor subunit 1 (VR1) [21]. Another research group prepared a library of 328,000 peptoid trimers, and used iterative deconvolution to discover micromolar ligands for both the melanocortin type 1 (MC1) and gastrin-releasing peptide/bombesin receptors [22]. Finally, a small family of peptoid trimers was generated in order to mimic the Agouti-related protein, including one member that showed micromolar affinity for the melanocortin type 4 receptor [23].

1.3.2
Peptoid Inhibitors of RNA-Protein Interactions

Peptoid libraries have also yielded compounds active in the disruption of RNA-protein interactions. Compounds not derived from library syntheses are discussed in Section 1.4.1. A peptoid 9-mer with a number of cationic groups was discovered (Fig. 1.4) after several rounds of mixture deconvolution, that was able to block the interaction of HIV-1 Tat protein with TAR RNA at nanomolar concentra-

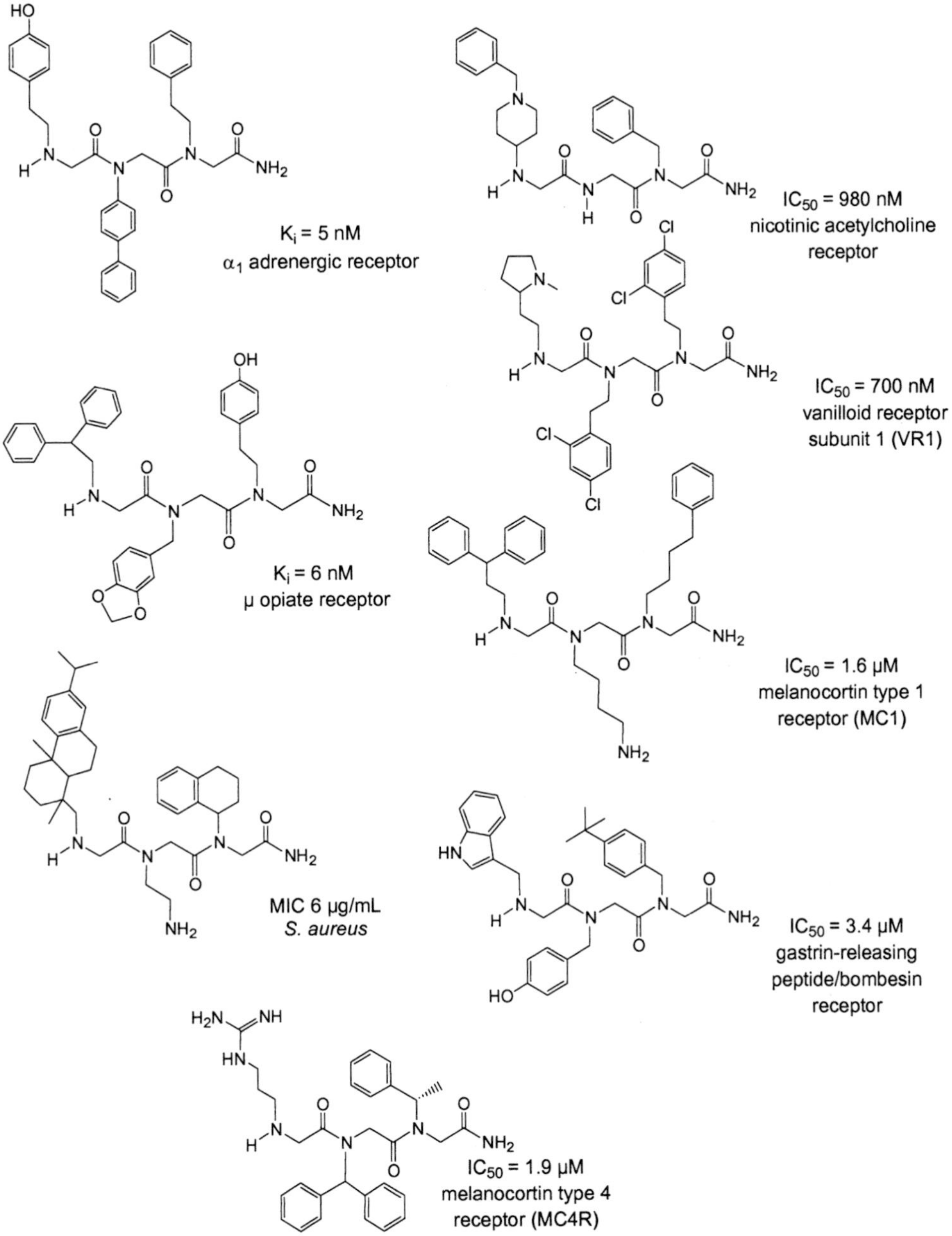

$K_i = 5$ nM
α_1 adrenergic receptor

$IC_{50} = 980$ nM
nicotinic acetylcholine
receptor

$IC_{50} = 700$ nM
vanilloid receptor
subunit 1 (VR1)

$K_i = 6$ nM
μ opiate receptor

$IC_{50} = 1.6$ μM
melanocortin type 1
receptor (MC1)

MIC 6 μg/mL
S. aureus

$IC_{50} = 3.4$ μM
gastrin-releasing
peptide/bombesin
receptor

$IC_{50} = 1.9$ μM
melanocortin type 4
receptor (MC4R)

Fig. 1.3 Various small-molecule peptoid ligands derived from combinatorial libraries

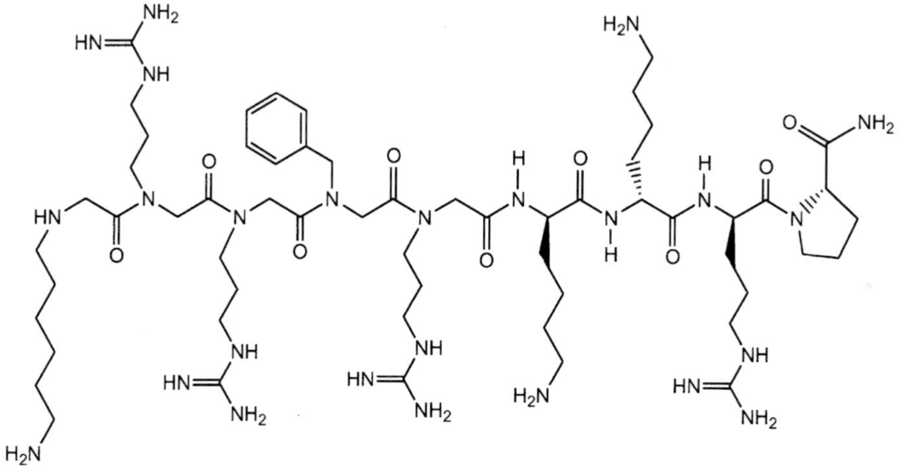

Fig. 1.4 Structure of a peptoid/peptide hybrid that is a submicromolar inhibitor of the HIV-1 Tat/Tar interaction

tions, as well as block HIV-1 replication *in vivo* [24]. Other cationic peptoids with similar side chains, which also inhibit this interaction have been discovered [25].

1.4
Peptoid-Based Drug Delivery and Molecular Transporters: Cellular Uptake

The cellular membrane presents a formidable barrier to drug uptake. In order to exhibit passive diffusion into the cell, drugs must be polar so as to facilitate distribution into the aqueous cellular environment, yet not so polar as to prevent diffusion across the hydrophobic interior of the cellular membrane. In addition, other physical drug characteristics (e.g. molecular weight $> \sim 700$ Da) can limit bioavailability. Promising drugs that do not possess the requisite characteristics for passive cellular entry can instead be delivered by novel techniques. For instance, a highly lipophilic drug might be delivered by packaging in liposomes, or a very polar drug might be functionalized with a lipophilic moiety. Similarly, certain large polycationic homopolymers of lysine [26], ornithine [27], and arginine [27] (between 4 and 200 kDa in weight) have been shown to facilitate membrane translocation in cells, and can be covalently ligated to biomolecules to promote cellular entry. However, these large polycations can be toxic *in vivo*, difficult to produce, and expensive.

1.4.1
Peptoid Mimics of HIV-Tat Protein

Alternatively, one interesting drug delivery technique exploits the active transport of certain naturally-occurring and relatively small biomacromolecules across the cellular membrane. For instance, the nuclear transcription activator protein (Tat) from HIV type 1 (HIV-1) is a 101-amino acid protein that must interact with a 59-base RNA stem–loop structure, called the *trans*-activation region (Tar) at the 5' end of all nascent HIV-1 mRNA molecules, in order for the virus to replicate. HIV-Tat is actively transported across the cell membrane, and localizes to the nucleus [28]. It has been found that the arginine-rich Tar-binding region of the Tat protein, residues 49–57 (Tat$_{49-57}$), is primarily responsible for this translocation activity [29].

Recently, there has been significant interest in peptidomimetic forms of Tat$_{49-57}$, not only because of its membrane translocation activity, but as a means of treating HIV infection [1]. Several peptoids, similar in sequence to Tat$_{49-57}$, have been synthesized with the intention of preventing the HIV-Tat/Tar interaction, and thus preventing HIV replication [24, 25, 30, 31]. However, only recently has this class of peptoids been applied to membrane translocation and drug delivery applications.

Short peptoid-based Tat$_{49-57}$ analogs are more advantageous drug delivery vehicles than large polycationic homopolymers of lysine, ornithine, and arginine. Not only are peptoids more readily synthesized and potentially bioavailable than such large polymers (Section 1.2), they are less likely to be proteolytically degraded [5] or to cause an immunological response than are peptide-based analogs.

Wender and colleagues investigated various truncated and alanine-substituted fluorescently labeled peptoid analogs of Tat$_{49-57}$ in order to determine the requisite structural features for membrane translocation activity in Jurkat cells [32]. They determined that the presence of at least six arginine residues (guanidino moieties) was critical for rapid cell entry, while no specific charge or structural element was important. Correspondingly, they synthesized a series of fluorescently-labeled oligoguanidine peptoids, and compared their capacity for cellular uptake in Jurkat cells with peptide D-arginine oligomers between five and nine residues in length. They identified one compound, *N*-hxg9, which was superior to a D-arginine nonamer in cellular uptake, which in turn was about 100-fold more rapidly translocated across the cell membrane than a fluorescently-labeled Tat$_{49-57}$.

1.4.2
Cellular Delivery of Nucleic Acids

Peptoids have also shown great utility in their ability to complex with and deliver nucleic acids to cells, a critical step toward the development of antisense drugs, DNA vaccines, or gene-based therapeutics. Most non-viral nucleic acid delivery systems are based on cationic molecules that can form complexes with the polyan-

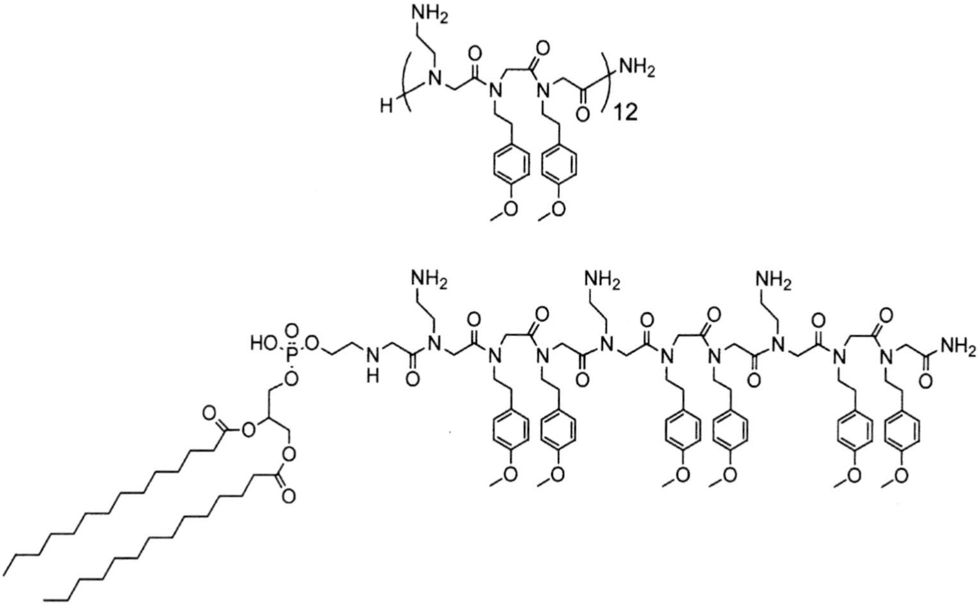

Fig. 1.5 Structure of a 36mer peptoid and the corresponding "lipitoid"

ionic nucleic acid [33]. These cationic materials are typically either polymeric or lipid-based structures, whose performance depends on a balance of factors including overall size, type of cation, density of charge, hydrophobicity, and solubility.

Peptoids are ideally suited to this task because their primary and secondary structures can be precisely predicted and controlled through their sequence. Combinatorial libraries of cationic peptoids were synthesized and evaluated for their ability to condense, protect, and deliver plasmid DNA to cells in culture [6]. An effective, 36mer compound was discovered that contained a repeating cationic trimer motif: cationic-hydrophobic-hydrophobic (Fig. 1.5). In an effort to refine the activity of this compound, solid-phase chemistry was developed to rapidly conjugate lipid moieties to the *N*-termini of peptoids. These cationic peptoid-lipid conjugates (called "lipitoids") were substantially more active in delivering plasmid DNA to cells, and also showed reduced cellular toxicity relative to the lead compound [34]. The most active lipitoids are composed of a natural phosphatidyl ethanolamine lipid conjugated to a 9mer with the same trimeric motif.

Although *in vivo* delivery studies using these reagents are still an early stage, high *in vitro* activity and low toxicity make lipitoids ideal transfection reagents. Recent work has shown that very small structural changes in the lipitoid can result in molecules that efficiently deliver antisense oligonucleotides and RNA.

1.5
Peptoid Mimics of Peptide Ligands

One straightforward approach to the design of biologically-active peptoid sequences is the systematic modification of an active peptide target sequence. For example, constituent amino acids of a target peptide may be substituted by peptoid residues with identical side chains. This modification results in a peptoid oligomer with side chains that are shifted relative to their original positions in the peptide template. An alternative approach is to generate a library of compounds in which site-specific substitutions are made using a diverse set of peptoid monomers, followed by screening to identify the most active compounds. In these methods, the partial substitution of amino acids by peptoid residues generates peptide/peptoid hybrids, which should ideally possess improved pharmokinetic and/or binding properties.

Characterization of peptoid-containing analogs of peptide ligands can provide valuable information. For example, this process can help to elucidate the position of critical residues in the protein target that provide important binding determinants. It can identify molecules with enhanced activity relative to the starting structure and/or with enhanced specificity to a protein target. Finally, it can identify active species with peptoid monomer substitution levels sufficient to grant significantly improved protease stability relative to the original peptide of interest.

The utility of peptoid/peptide hybrids was exemplified by a study of hybrid ligands to proteins containing the Src homology 3 (SH3) domain [35, 36]. As part of their signal transduction activity, these proteins recognize peptides with a PxxP motif (where P = proline, and x = any other amino acid). In general, peptide ligand binding to SH3 domains occurs with low affinity and low specificity. Peptide sequences derived from wild-type protein partners exhibit significant binding to the large family of proteins with SH3 domains. Nguyen *et al.* conducted studies aimed at an improved understanding of the role of proline residues in the core PxxP motif at the binding interface. Previous structural studies had suggested that recognition requires the presence of an *N*-substituted amino acid at proline positions, but that proline itself was not specifically required. Peptide/peptoid hybrids containing a variety of different peptoid substitutions for proline in the core PxxP region were synthesized and evaluated for binding to various proteins containing SH3 domains. It was found that single-site substitutions can have a strong effect on binding affinity. For instance, a hybrid oligomer bearing a dimethoxybenyzl *N*-substitution showed a 100-fold increase in binding affinity over the wild-type peptide (to a K_D of 30 nm), accompanied by a dramatic gain in specificity for binding to the SH3 domain of the protein *N*-Grb2 relative to those of Src and Crk. Multiple substitutions were also well tolerated. For example, an analog of a peptide with partial specificity for Crk was synthesized, in which both proline residues in the PxxP motif were replaced by peptoid monomers. The new hybrid retained strong binding to Crk, but no longer recognized Src or *N*-Grb2 [35]. The Goodman group has similarly investigated the effect of a variety of peptoid substitutions at the proline position in a cyclic hexapeptide analog of somatostatin, cyclo-

[Phe-Pro-Phe-D-Trp-Lys-Thr]. This study also yielded compounds with an enhanced binding selectivity, in this case, for the hsst2 receptor [37].

An alternative approach involved the use of a "peptoid scan" strategy to identify peptoid–peptide hybrids that bind to the Src homology 2 (SH2) domain of Syk tyrosine kinase. Ruitjenbeek *et al.* investigated hybrid oligomers in which conserved side chains of the Syk ligand Ac-pTyr-Glu-Thr-Leu-NH$_2$ were shifted sequentially to the corresponding *N*-pendant peptoid positions [38]. They found that peptoid substituents were accommodated at one or both of the Thr and Leu positions with less than 10-fold loss of inhibitory activity and binding affinity but that substitutions at other positions eliminated activity. These results were ascribed to the alteration of distances between side chains and/or the loss of hydrogen-bond contacts between the *N*-terminal region of the ligand and the protein target.

The Liskamp group also examined the ability of peptoid–peptide hybrids to be bound by the MHC Class II receptor, an important component of the human immune system [39]. Two of three peptoid substitutions in the 14-residue peptide caused substantial decreases in binding affinity, despite the fact that these were solvent-exposed residues. These results were attributed to a loss of hydrogen-bond contacts as well as to steric clashes caused by unfavorable positioning of the new side chain groups.

The Kessler group has generated side chain-shifted peptoid substitutions within peptidic antagonists of the $\alpha 4\beta 7$ integrin interaction with the mucosal addressin cell adhesion molecule 1 (MAdCAM-1) [40]. Peptoid substitutions within the core Leu-Asp-Thr sequence of the peptide abolished activity in a cell adhesion assay. However, it was possible to substitute the phenylalanine in the cyclo(Phe-Leu-Asp-Thr-Asp-D-Pro) peptide with analogous peptoid and azapeptide monomers, with some retention of activity. The extreme sensitivity of activity to substitutions in the core sequence was attributed to a need to maintain a precise spacing between side chains, and to the importance of the peptide backbone, in this case, in providing productive binding interactions.

An evaluation of the several attempts to find peptoid analogs for wild-type peptide ligands suggests that it is possible, in some cases, to identify specific positions amenable to peptoid substitution. In particular, sites at which binding occurs through recognition of proline residues and in which hydrogen bonding does not play a strong role may be particularly suitable for substitution by peptoid residues. However, it may be challenging to fully replace protein-binding peptide sequences entirely by peptoids. Difficulties with the "translation" of peptides into peptoid sequences may arise due to some of the following reasons:

1. In cases where the presentation of side-chain functionalities and backbone hydrogen bonding partners with a precise spacing and geometry are essential for optimal binding interactions in the wild-type peptide, peptoids will exhibit reduced binding efficacy due to the lack of amide NH groups and the modified distances between side chains and adjacent carbonyls. Some attempts have been made to address this issue by the synthesis of retropeptoid sequences, with mixed results [7].

2. If peptide residues are converted to peptoid residues, the conformational heterogeneity of the polymer backbone is likely to increase due to *cis/trans* isomerization at amide bonds. This will lead to an enhanced loss of conformational entropy upon peptoid/protein association, which could adversely affect binding thermodynamics. A potential solution is the judicious placement of bulky peptoid side chains that constrain backbone dihedral angles.

3. If the individual residues of a peptide are transformed into the corresponding peptoid monomers to make hybrid oligomers, there will be a perturbation in the distance between side chains at the boundary between oligomer types. That is, spacing of side chains at a peptoid-peptide linkage will be different from that between either two peptide or peptoid residues.

1.6
Peptoids with Folded Structure

One of the most intriguing features of peptoids is their capacity to adopt stable, helical conformations in a variety of solvent media. The helix is an important component of folded protein structure. Additionally, helical polypeptides have important roles *in vivo* (e.g. in certain ligand-binding interactions and in the transmembrane regions of membrane-bound proteins). As such, similar peptoid structures may be valuable tools for drug development. Due to their ability to adopt folded structure, peptoids are among a select class of sequence-specific peptidomimetic materials, referred to as "foldamers" [41], which can be designed to adopt specific backbone conformations. This section describes early work in the discovery and characterization of helical peptoid structures. The following section (Section 1.7) then details some recent applications of these peptoid helices.

1.6.1
Restricting Conformational Space

The inherently achiral nature of the oligo-(N-substituted) glycine backbone enables peptoid oligomers with achiral, unbranched side chains to sample a large conformational space. However, experimental work has shown that the incorporation of side chains with branched carbons immediately adjacent to the main-chain nitrogen to which they are appended (*a*-branched) allows the formation of helices with *cis*-amide bonds [4]. Modeling of a peptoid octamer with this type of side chain confirmed that substitution with bulky, chiral groups significantly reduces the energetically accessible conformational space [3]. As N-substituted glycines are incapable of backbone hydrogen bonding, structures in these conformationally restricted peptoids were instead predicted to be stabilized by a combination of side chain–backbone steric repulsions and dipole–dipole repulsions between main

Tab. 1.2 Examples of chiral α-methyl N-substituted glycine side chains

	N-substituted glycine oligomer, or *peptoid*

R = Side chain	Designator
	Nsch = (S)-N-(1-cyclohexylethyl)glycine
	Nssb = (S)-N-(2-butyl)glycine)
	Nspe = (S)-N-(1-phenylethyl)glycine

chain amide carbonyl electrons and, for specific peptoid sequences, aromatic side chains [3]. It was later determined that the incorporation of bulky, chiral, α-methyl side chains can provide significant conformational restriction without compromising synthetic yields (Tab. 1.2) [4].

1.6.2
Peptoid Helices

1.6.2.1 CD and NMR Studies of a Helical Peptoid Pentamer with α-Chiral Aromatic Side Chains

There have been several studies that investigated, by a variety of techniques, the folding behavior of peptoids with diverse chiral, α-branched side chains (Tab. 1.2) [4, 42]. Peptoids as short as five monomers in length, substituted with chiral aromatic residues, exhibit circular dichroism (CD) spectra indicative of chiral secondary structure, with spectral features similar to those of peptide α-helices [42]. Structural handedness is dictated by the choice of side-chain enantiomer, such that peptoids containing side chains of opposing handedness exhibit mirror-image CD (Fig. 1.6) [4, 43]. The CD spectra of homooligomeric peptoid sequences based on the α-chiral, aromatic monomer (S)-N-(1-phenylethyl)glycine are particularly intense, with a maximum near 190 nm and double minima near 204 and 218 nm

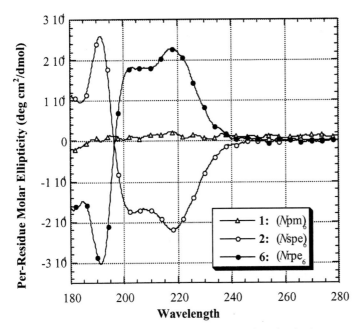

Fig. 1.6 A comparison of the CD spectra of oligopeptoids with achiral Npm side chains (1) and with α-chiral, aromatic sidechains of S and R chirality (2 and 6, respectively). Sample concentration was ~60 μM in acetonitrile. Spectra were acquired at room temperature. Npm = (N-[1-phenylmethyl]glycine); Nspe = (S)-N-(1-phenylethyl)glycine; Nrpe = (R)-N-(1-phenylethyl)glycine

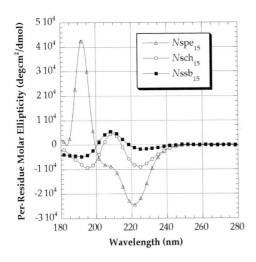

Fig. 1.7 Illustration of the distinct CD exhibited by peptoid helices containing solely aromatic or solely aliphatic residues. Sample concentrations were ~60 μM in acetonitrile. Spectra were acquired at room temperature. Nspe = (S)-N-(1-phenylethyl)glycine; Nsch = (S)-N-(1-cyclohexylethyl)glycine; Nssb = (S)-N-(sec-butyl)glycine

[4, 42, 44] (Figs. 1.6 and 1.7). This type of CD spectrum is observed for certain heterooligomeric peptoid sequences with as few as 33% chiral aromatic residues, in both aqueous and polar organic solvent (acetonitrile, methanol).

A peptoid pentamer of five *para*-substituted (S)-*N*-(1-phenylethyl)glycine monomers, which exhibits the characteristic *a*-helix-like CD spectrum described above, was further analyzed by 2D-NMR [42]. Although this pentamer has a dynamic structure and adopts a family of conformations in methanol solution, 50–60% of the population exists as a right-handed helical conformer, containing all *cis*-amide bonds (in agreement with modeling studies [3]), with about three residues per turn and a pitch of ∼6 Å. Minor families of conformational isomers arise from *cis/trans*-amide bond isomerization. Since many peptoid sequences with chiral aromatic side chains share similar CD characteristics with this helical pentamer, the type of CD spectrum described above can be considered to be indicative of the formation of this class of peptoid helix in general.

1.6.2.2 CD Studies of Longer Peptoid Helices Containing *a*-Chiral Aromatic Side Chains

A more comprehensive folding study by CD and NMR that investigated aromatic side chain-containing peptoids in acetonitrile solution was carried out subsequent to the NMR structural studies described above. In particular, the effect of chain length on structure in homooligomers of (R)-*N*-(1-phenylethyl)glycine (Nrpe) [44] and the effect of sequence on structure in a variety of more than 30 heterooligomeric peptoids were studied in acetonitrile solution [45]. CD spectra of homooligomers of Nrpe display helical characteristics at lengths as short as five residues, in agreement with the modeling results of Armand and co-workers [3] and prior observations described above [4]. The CD spectral intensity initially diminishes as peptoid chain length increases from five to about eight residues, with spectra passing through an isodichroic point at 196 nm. At a chain length of nine monomers, a unique, non-helical CD spectrum is observed (in acetonitrile solution), which we have since found corresponds to a novel planar cyclic peptoid structure, the details of which are described elsewhere [46]. At the 10-monomer length, the CD spectrum is again characteristically helical, and gradually intensifies with increasing chain length. At a length of 13 residues and longer, the spectrum is fully developed, and exhibits no further intensification or shape change as the chain is further lengthened through 20 monomers [44].

General sequence requirements for helix formation in heterooligomeric peptoids [45] have also been determined. Peptoids ranging in length from six to 24 residues and containing mixed chiral and non-chiral aromatic and aliphatic side chains were synthesized and studied by CD and 2D-NMR. Although the prior studies of Kirshenbaum and colleagues [4] suggested that helical sequences contained at least two-thirds *a*-chiral residues, with at least one-third also being *a*-chiral aromatic, Wu *et al.* endeavored to determine sequence requirements for helix formation in peptoids with non-periodic sequences. They found peptoids which include achiral residues form stable helices in sequences as short as six residues if at least 50% *a*-chiral aromatic residues are incorporated. In peptoids shorter than about 12–15 residues, helix formation is also promoted by incorporation of an *a*-chiral aromatic residue at the peptoid *C*-terminus, as well as by inclusion of an aromatic helical face (patterned via three-fold sequence periodicity) and a carboxyamide *C*-terminal moiety. In longer peptoids (>12–15 residues), residue placement effects seem less important, although the percentage of *a*-chiral aromatic side chains required for helicity remains unchanged.

In general, peptoid helical structure is thermally stable in both organic solvent and in aqueous solution [4, 44, 47]. Complete thermal unfolding of peptoid helices has so far not been observed [44, 47]. However, studies of several different peptoid oligomers have shown a diminishing CD signal with increasing temperature, with spectra passing through one or more isodichroic points. These observations indicate that temperature can be used to induce partial melting of repeating peptoid secondary structure. Although it was reported that a water-soluble peptoid 30mer underwent a cooperative and reversible unfolding transition due to combined thermal and pH effects [4], we found with further studies of this molecule that the observed phase transition was complicated by a solubility-to-insolubility, LCST-type phase transition (unpublished results).

The thermal and solvent stability of aromatic side chain-containing peptoid helices was investigated in more detail by Sanborn and colleagues using CD in a heterooligomeric 36mer sequence, containing one-third (S)-*N*-(1-phenylethyl)glycine (*N*spe) residues and two-thirds chiral, anionic aliphatic (S)-*N*-(1-carboxyethyl)-glycine (*N*sce) residues [47]. Aromatic side chains were patterned with 3-fold periodicity in this sequence. This peptoid displayed the concentration-independent *a*-helix-like CD spectrum that is characteristic of the aromatic side chain-containing peptoid helices previously described by Kirshenbaum *et al.* [4] and Wu *et al.* [44, 45]. Sanborn *et al.* found that the folded structure of this peptoid 36mer, as reported by CD, was remarkably robust. The helical CD signal is invariant over an increase of three orders of magnitude in buffer concentration, and is stable to treatment with 8 M urea and to heating through 75 °C (separately, and in combination). The extraordinary resilience of this peptoid helix to unfolding agrees with the dominance of steric forces in helix stabilization, consistent with prior hypotheses [3]. Correspondingly, the CD signal of this 36mer peptoid is also largely invariant with solvent (water, acetonitrile, and 20 mol% aqueous 2,2,2-trifluoroethanol), showing modest intensification of the CD signal in aqueous TFE (in a manner analogous to polypeptides [48]).

1.6.2.3 Structural Studies of Peptoids with Aliphatic Side Chains by CD, NMR, and X-ray Crystallography

Peptoids based on *a*-chiral aliphatic side chains can form stable helices as well [43]. A crystal of a pentameric peptoid homooligomer composed of homochiral *N*-(1-cyclohexylethyl)glycine residues was grown by slow evaporation from methanol solution, and its structure determined by X-ray crystallographic methods. In the crystalline state, this pentamer adopts a helical conformation with repeating *cis*-

amide bonds, a periodicity of ~3 residues per turn, and a pitch of ~6.7 Å. Longer peptoid homooligomers of *N*-(1-cyclohexylethyl)glycine (12–15 residues) give rise to fully developed and intense CD signals, distinct from those obtained from aromatic side chain-containing peptoids (Fig. 1.7). In contrast, the CD spectrum of this aliphatic peptoid is strongly reminiscent of the spectrum of a peptide polyproline type-I helix, and contains a distinct maximum at 210 nm and two shallow minima at 200 and 225 nm. 2D-NMR studies of 6, 9, 12, and 15mer peptoid homo-oligomers of *N*-(1-cyclohexylethyl)glycine were also undertaken [43]. When considered jointly, the highly degenerate NMR spectra (which are increasingly degenerate at longer chain lengths), CD spectra that intensify as peptoid chain length increases, along with the helical crystal structure described above, corroborate the adoption of a repeating helical structure similar to that observed in the *N*-(1-phenylethyl)glycine pentamer previously studied by 2D-NMR [42]. These NMR spectra indicate that there are two major families of conformers present in solution, specifically *cis–trans* isomers of the backbone amide bonds present in a 2.6 : 1 ratio respectively, in acetonitrile solution.

1.6.2.4 Summary

Appropriately substituted peptoids longer than about five residues can form highly stable helical structures. Sequences that contain at least 50% *a*-chiral aromatic residues can form stable helices, the handedness of which may be controlled by choice of side chain enantiomer. The extent of helical structure increases as chain length grows, and for these oligomers becomes fully developed at a length of approximately 13 residues. Aromatic side chain-containing peptoid helices generally give rise to CD spectra that are strongly reminiscent of that of a peptide *a*-helix, while peptoid helices based on aliphatic groups give rise to a CD spectrum that resembles the polyproline type-I helical CD. Despite the differences in CD spectra, both peptoid types form the same type of helix, which is structurally similar to the type-I polyproline helix. As peptoid helices are stabilized predominantly by steric repulsions, their structure is remarkably stable to thermal, ionic, and chaotropic destabilizing influences.

1.6.3
Protein-mimetic Structures

Proteins derive their powerful and diverse capacity for molecular recognition and catalysis from their ability to fold into defined secondary and tertiary structures and display specific functional groups at precise locations in space. Functional protein domains are typically 50–200 residues in length and utilize a specific sequence of side chains to encode folded structures that have a compact hydrophobic core and a hydrophilic surface. Mimicry of protein structure and function by non-natural oligomers such as peptoids will not only require the synthesis of >50mers with a variety of side chains, but will also require these non-natural sequences to adopt, in water, tertiary structures that are rich in secondary structure.

In order for folded helices to assemble into tertiary structures in water, they need to be amphipathic (e.g. where one helical face is hydrophobic and the other is hydrophilic). Because the first helical peptoids contained very hydrophobic chiral residues, ways to increase the water solubility and side-chain diversity of the helix-inducing residues were investigated [49]. It was found that a series of side chains with chiral-substituted carboxamides in place of the aromatic group could still favor helix formation, while dramatically increasing water solubility.

Because there is little precedent for the *de novo* design and synthesis of a completely synthetic macromolecule of defined structure [50], a combinatorial synthesis and screening process was used to identify sequences that could adopt tertiary structures [14]. Rather than attempt to synthesize a continuous single chain, discrete single amphipathic 15mer oligomers were made and tested for their ability to assemble into defined multimers. This was accomplished in a high-throughput mode by adding a dye molecule that only fluoresces when bound in a hydrophobic environment – such as a folded protein core. A library of 3400 water-soluble amphiphilic helices was screened for dye binding and a small number of 15mers were discovered that formed defined helical assemblies as judged by size-exclusion chromatography, circular dichroism and analytical ultracentrifugation [14]. This is a significant initial step toward the synthesis of an artificial protein, but in itself does not yet represent true tertiary structure. Future work will investigate the formation of such bundles in longer, contiguous peptoid oligomers.

1.7
Biomimetic Peptoid Structures for Therapeutic Applications

The well-defined helical structure associated with appropriately substituted peptoid oligomers (Section 1.6) can be employed to fashion compounds that closely mimic the structure and function of certain bioactive peptides. There are many examples of small helical peptides (< 100 residues) whose mimicry by non-natural oligomers could potentially yield valuable therapeutic and bioactive compounds. This section describes peptoids that have been rationally designed as mimics of antibacterial peptides, lung surfactant proteins, and collagen proteins. Mimics of HIV-Tat protein, although relevant to this discussion, were described previously in this chapter (Sections 1.3.2 and 1.4.1).

1.7.1
Peptoid Mimics of Antibacterial Peptides

The magainins are a class of linear, cationic, facially amphipathic and helical antibacterial peptides derived from frog skin [51]. The magainins exhibit highly selective and potent antimicrobial activity against a broad spectrum of organisms [52, 53]. As these peptides are facially amphipathic, the magainins have a cationic helical face (at physiological pH), composed of mostly lysine residues, as well as hydrophobic aromatic (phenylalanine) and hydrophobic aliphatic (mostly valine, leucine, and isoleucine) helical faces. It has been shown that the structure and physicochemical properties of the magainins, rather than any specific receptor–ligand interactions, are responsible for their activity [54].

As such, the magainins provide a useful initial target for peptoid-based peptidomimetic efforts. Since the helical structure and sequence patterning of these peptides seem primarily responsible for their antibacterial activity and specificity, it is conceivable that an appropriately designed, non-peptide helix should be capable of these same activities. As previously described (Section 1.6.2), peptoids have been shown to form remarkably stable helices, with physical characteristics similar to those of peptide polyproline type-I helices (e.g. *cis*-amide bonds, three residues per helical turn, and ~6 Å pitch). A facially amphipathic peptoid helix design, based on the magainin structural motif, would therefore incorporate cationic residues, hydrophobic aromatic residues, and hydrophobic aliphathic residues with threefold sequence periodicity.

A series of peptoid magainin mimics with this type of three-residue periodic sequence, has been synthesized [55]. Each was purified by reversed-phase HPLC to >97%, and their minimum inhibitory concentrations (MIC) determined (using broth-dilution techniques) against both *E. coli* JM109 and *B. subtilis* BR151 in Luria Broth medium (Tab. 1.3). The peptoids in Tab. 1.3 have been arranged in order of increasing hydrophobicity, as determined from reversed-phase HPLC retention times. The T2 and T3 sequence motifs, at lengths between 12 and 17 residues, are the most effective antibacterial compounds, with low micromolar and even submicromolar MIC values, similar to those of magainin-2 amide. In all cases, peptoids are individually more active against the Gram-positive species.

Tab. 1.3 Magainin-mimetic peptoid sequences, and antibacterial and hemolytic activities

Peptoid	Sequence	MIC (µM)		% hemolysis
		E. coli JM109	B. subtilis BR151	At E. coli JM109 MIC
T1-17	H-Nssb-Nssb-(NLys-Nssb-Nssb)$_5$-NH$_2$	>100	>100	0%
T2-6	H-(NLys-Nssb-Nspe)$_2$-NH$_2$	>487	>730	0%
T2-9	H-(NLys-Nssb-Nspe)$_3$-NH$_2$	218	55–82	0%
T2-12	H-(NLys-Nssb-Nspe)$_4$-NH$_2$	49	7.8	0%
T2-15	H-(NLys-Nssb-Nspe)$_5$-NH$_2$	9.9	4.4	0%
T2-17	H-Nspe-Nspe-(NLys-Nssb-Nspe)$_5$-NH$_2$	19	1.4	1.2%
T3-12	H-(NLys-Nspe-Nspe)$_4$-NH$_2$	9.9	0.82	1.4%
T3-17	H-Nspe-Nspe-(NLys-Nspe-Nspe)$_5$-NH$_2$	7.7	1.2	51%
T4-17	H-Nsch-Nsch-(NLys-Nsch-Nsch)$_5$-NH$_2$	>75	>75	100%

Nssb = (S)-*N*-(sec-butyl)glycine; Nspe = (S)-*N*-(1-phenylethyl)glycine; Nsch = (S)-*N*-(1-cyclohexylethyl)glycine; NLys = *N*-(4-aminobutyl)glycine.

Certain of these peptoid antibiotics are also selective for bacterial, rather than mammalian, cells. The selectivity of these peptoids has been measured in terms of their capacity to cause hemolysis of human erythrocytes at or near their MIC (Tab. 1.3). Interestingly, the amount of hemolysis induced by these peptoids correlates well with their hydrophobicity, as there is an increasing extent of hemolysis as molecular hydrophobicity increases. These results suggest that highly hydrophobic compounds of this class are poorly selective antibiotics. The most active antibacterial peptoids, T2-15 and T3-12, have quite low hemolytic activity near their MICs. Although highly antibacterial *in vitro*, T3-17 is also very hemolytic at its MIC value.

All of these water-soluble peptoids were designed to be helical, based on the loose guidelines provided by previous studies (Section 1.6.2). In some cases, though, these were novel sequences whose helical extent could not be predicted. Therefore, circular dichroism spectroscopy was used to characterize the structure of these oligomers in solution. In aqueous Tris-HCl buffer (pH 7.0), the most effective peptoids (T2 and T3 between 12 and 17 residues long) exhibit a spectrum characteristic of the peptoid helix, with a single ellipticity maximum near 190 nm, and double ellipticity minima near 202 and 220 nm. Interestingly, ineffective antibacterial compounds T1-17 and T4-17 both exhibit a random coil-like CD in both aqueous buffer and hydrophobic solvent environments [55]. We also find that selective (non-hemolytic) and effective antibacterial peptoids exhibit intense, helical CD in bacterial-mimetic vesicles (composed of 70 mol% POPE : 30 mol% POPG), but show relatively weak CD in mammalian red blood cell-mimetic DMPC vesicles, nearly indistinguishable from CD in aqueous buffer. On the other hand, non-selective (hemolytic) antibacterial peptoids exhibit strong CD intensification in both types of vesicles, as compared to CD in buffer [55].

In summary, these recently obtained results demonstrate that certain amphipathic peptoid sequences designed to mimic both the helical structure and approximate length of magainin helices are also capable of selective and biomimetic antibacterial activity. These antibacterial peptoids are helical in both aqueous buffer and in the presence of lipid vesicles. Ineffective (non-antibacterial) peptoids exhibit weak, random coil-like CD, with no spectral intensification in the presence of lipid vesicles. Selective peptoids exhibit stronger CD signals in bacterial-mimetic vesicles than in mammalian-mimetic vesicles. Non-selective peptoids exhibit intensely helical CD in both types of vesicles.

1.7.2
Peptoid-Based Mimics of Lung Surfactant Proteins

There is a clinical need for non-natural, functional mimics of the lung surfactant (LS) proteins B and C (SP-B and SP-C), which could be used in a biomimetic LS replacement to treat respiratory distress syndrome (RDS) in premature infants [56]. An effective surfactant replacement must meet the following performance requirements: (i) rapid adsorption to the air–liquid interface, (ii) re-spreadability upon multiple, successive surface compressions, and (iii) attainment of near-zero surface tension upon compression [57]. Currently, neonates suffering from RDS are treated with bovine-derived surfactant. However, these surfactant preparations are expensive and may contain viruses or other hazards, which raises concerns regarding the safety of their clinical administration. Entirely synthetic surfactants, which do not contain SP-B and SP-C, currently do not meet the performance requirements outlined above [58]. As an alternative, in recent work peptoids have been used to create functional mimics of both SP-B and SP-C with the eventual goal of creating better synthetic LS replacements [59].

SP-C is a helical, cationic, amphipathic protein just 35 amino acids in length. Its structure is relatively simple, containing a hydrophobic α-helix 37 Å long, capped by two adjacent, positively-charged residues, and a relatively unstructured amino terminus [60, 61]. The valyl-rich helix of SP-C inserts into lipid monolayers and bilayers and interacts with hydrophobic alkyl tails, while cationic side chains interact with polar lipid head groups. This ionic interaction is thought to bind SP-C to the lipid monolayer [60, 62]. In contrast, SP-B has a more complex structure, which is also predominantly helical, but with numerous intra-chain disulfide bonds. *In vivo*, the SP-B protein exists as a homodimer via an additional disulfide linkage. However, a short, helical segment from the *N*-terminus of one monomer (SP-B$_{1-25}$) seems to retain much of the biologically relevant surface activity of the full-length synthetic peptide [63, 64]. CD studies show that SP-B$_{1-25}$ is α-helical in structure. NMR studies, as well as modeling and sequence analysis, predict that this helix has both cationic and non-polar helical faces [65, 66].

Recently, peptoid-based mimics of both SP-C and SP-B have been designed to adopt helical secondary structures, and also mimic (to varying degrees) the sequence patterning of hydrophobic and polar residues found in the natural surfactant proteins. Peptoid-based SP-C mimics of up to 22 monomers in length, were synthesized and characterized by *in vitro* experimental methods [67, 68] (Fig. 1.8). The secondary structure of all molecules was assessed by circular dichroism and found to be helical. The surface activities of these peptoids, in comparison to the actual SP peptides described above, were characterized by surfactometry using

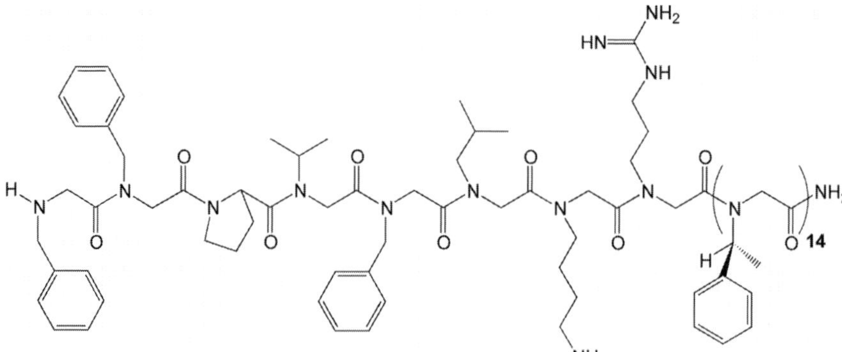

Fig. 1.8 Structure of a peptoid mimic of SP-C$_{5-32}$

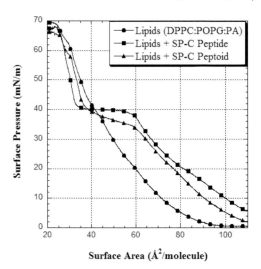

Fig. 1.9 Surface pressure (π)–area (A) isotherms obtained for a lipid mixture (DPPC:POPG:PA, 68:22:9 (by weight)), alone and with 10% (w/w) of either SP-C peptide or SP-C peptoid added. Results indicate that the addition of the SP-C mimics engenders biomimetic surface activity, as indicated by lift-off at a higher molecular area and the introduction of a plateau

Legend:
— Lipids (DPPC:POPG:PA)
— Lipids + SP-C Peptide
— Lipids + SP-C Peptoid

Y-axis: Surface Pressure (mN/m)
X-axis: Surface Area (Å²/molecule)

both a Langmuir–Wilhelmy surface balance (LWSB) in conjunction with fluorescence microscopy (FM), and a pulsating bubble surfactometer (PBS).

When added to a biomimetic lipid mixture (DPPC:POPG:PA, 68:22:9, which behaves similarly to the natural lipid mixture found in pulmonary surfactant), a variety of peptoid SP analogs show biomimetic surface activity and a substantial improvement over that which is observed for a lipid film without SP-mimics. Surfactometry results in general, show peptoid SP-mimics behave similarly to the natural SP-C peptide and the SP-B$_{1-25}$ mimic. Specifically, LWSB experiments reveal high lift-off surface areas (i.e. rapid surface adsorption) and high film-collapse surface pressures, as well as a plateau in the π-A isotherm that is characteristic of natural LS (Fig. 1.9). Surface phase morphology, observed by LWSB in conjunction with FM, shows that the SP-mimics cause biomimetic fluidization of the surface film at high compression. PBS studies reveal dynamic surface activity similar to that of the analogous α-peptides, specifically an increased rate of surface adsorption and decreased minimum and maximum surface tensions at the minimum and maximum bubble surface areas, respectively. Hence, peptoid mimics of the surfactant proteins seem to be promising as biostable replacements for SP-B and SP-C, and may prove useful in the development of novel therapies for the treatment of respiratory distress. They also provide an example, similar to the magainin-like peptoids, of structural mimicry enabling good recapitulation of function in a membrane-interactive oligomer.

1.7.3
Collagen-based Structures Containing Peptoid Residues

Collagen is the major protein component of connective tissue and constitutes approximately 25% of the total protein content in humans. There are more than 19 different types of collagen known to exist. The most abundant type, Type I, is composed of ~ 300 repeats of a three-amino acid sequence, (Gly-Xaa-Yaa), in which Xaa and Yaa are most commonly proline (Pro) and hydroxyproline (Hyp) respectively. In the native conformation, three of these long peptide chains adopt left-handed polyproline type-II-like helices, which are in turn twisted about one another into a right-handed helical superstructure. This helix, created by entwining three individual helices together, is called the collagen triple helix. These triple helices are further organized into fibrils, which provide collagen with its mechanical strength [69].

Collagen-like molecules have been synthesized which incorporate peptoid residues and which are capable of forming a triple helix structure. These peptoid-peptide chimerae are hypothesized to be more biostable and more resistant to enzymatic proteolytic degradation than natural collagen. Such resilient and durable collagen constructs would conceivably find numerous applications as biomaterials. Toward this end, Goodman and co-workers have published several reports in which they have investigated substitution of either or both Xaa and Yaa positions within the repetitive trimeric collagen sequence by various peptoid residues.

Initially, Goodman's group studied structures with the repetitive trimeric sequence (Gly-Pro-Nleu)$_n$ ($n=1$, 3, 6, 9) in which the C-terminus was amidated and the N-terminus was either acetylated or covalently ligated to Kemp triacid (KTA), a rigid trivalent template that promotes triple helix formation [70]. Here, Nleu signifies the N-isobutylglycine peptoid residue. Through CD, NMR, and molecular modeling experiments, Goodman and colleagues found that KTA-[Gly-(Gly-Pro-Nleu)$_6$-NH$_2$]$_3$ forms a collagen-like triple helix structure [71]. Moreover, this substitution of Nleu into the collagen sequence did not seem to significantly reduce helical stability. In fact, KTA-[Gly-(Gly-Pro-Nleu)$_6$-NH$_2$]$_3$ has a melting temperature somewhat higher than Ac-(Gly-Pro-Pro)$_{10}$ [72].

Other collagen-like structures that incorporate the repetitive trimeric sequence (Gly-Nleu-Pro) were found to be even more stable in the triple-helix conformation than analogous (Gly-Pro-Nleu) structures [73, 74]. Certain "host–guest" structures, in which an achiral (Gly-Nleu-Nleu) sequence was incorporated in between collagen-like repeats of (Gly-Pro-Hyp), were synthesized and retained a triple-helix conformation [75]. Additional "host–guest" constructs were created in which the (Gly-Nleu-Nleu) sequence was located in between repeats of (Gly-Nleu-Pro) [76] and were also found to adopt the triple helix conformation. However, these host–guest structures were significantly less stable than analogous constructs composed solely of (Gly-Pro-Hyp) or (Gly-Nleu-Pro) repeats.

The ability of these peptidomimetic collagen-structures to adopt triple helices portends the development of highly stable biocompatible materials with collagen-like properties. For instance, it has been found that surface-immobilized (Gly-Pro-Nleu)$_{10}$-Gly-Pro-NH$_2$ in its triple-helix conformation stimulated attachment and growth of epithelial cells and fibroblasts *in vitro* [77]. As a result, one can easily foresee future implementations of biostable collagen mimics such as these, in tissue engineering and for the fabrication of biomedical devices.

1.8
Obstacles to the Development of Biomedically-useful Peptoids

Numerous barriers remain to be overcome before the promise of biomedically-useful peptoids can be more completely fulfilled. The following sections detail some examples.

1.8.1
Enhance the Diversity of Secondary Structure in Peptoid Foldamers

Although high-resolution structures have been obtained of the polyproline type-I helix formed by certain oligopeptoids, circular dichroism spectra suggest that distinct sequence motifs can form alternative types of secondary structure (Section 1.6). Ongoing studies seek new side chain types that will generate novel, stable folded species. One outstanding challenge in particular is the quest for monomers that will provide for side-chain/main-chain and side-chain/side-chain hydrogen bonding [49].

1.8.2
Improve Understanding of Peptoid Sequence/Structure Relationships

Secondary structure prediction from the primary sequence of polypeptides has improved continuously over the course of more than 30 years. This approach has allowed the *a priori* design of peptides that present chemical functionalities in a predetermined orientation. Structure prediction tools have been pivotal for guiding molecular recognition strategies in peptides, and as we gain a more detailed knowledge of structure types formed by polypeptoids, these tools may also prove to be of similar utility in these non-natural systems. Peptoids have been shown to be amenable to structural prediction techniques, since the polyproline type-I helix formed by peptoid oligomers containing chiral N-(1-phenylethyl)glycine residues was predicted by a combined molecular mechanics/quantum mechanical approach [3].

1.8.3
Translate Bioactive Peptide Sequences into Bioactive Peptoid Sequences

There have been notable successes in the replacement of individual peptide residues by peptoid monomers with retention of *in vitro* activity and enhancement of specificity. Unfortunately, attempts to completely transform those bioactive peptides that function via specific peptide–protein binding events into entirely peptoid-based oligomers have so far proven successful only at short chain lengths (e.g. [23]). It remains to be seen whether any general strategy can be developed in order to address this difficulty, either for longer peptoid oligomers or in other peptidomimetic oligomers. Again, computational tools may prove essential. On the other hand, substantial success has been achieved in the mimicry of membrane-active peptides using peptoid sequences, including peptoid molecular transporters, antimicrobial peptoids, and SP-mimetic peptoids.

1.8.4
Develop Peptoids with Stable Tertiary Structure

The "foldamer" field is still in a juvenile stage regarding the mimicry of diverse peptide secondary structures [78]. Nevertheless, it is not too soon to address challenges in the mimicry of stable protein folds. The ability to design true proteomimetics that incorporate tertiary folds will eventually enable a more sophisticated approach to molecular recognition problems and may ultimately lead to non-natural enzyme-like molecules capable of catalysis and regulation. Preliminary approaches toward non-natural tertiary structure are likely to focus on identifying and exploiting folding forces driven by the solvophobic sequestration of polymer side chains [14]. An important prerequisite will be improving synthetic yields so that polypeptoids of sufficient length to form a solvent-shielded core can be made readily. Although high yields are a characteristic of the solid-phase sub-monomer peptoid synthetic strategy, the method will require further refinement in order to further improve the yield of long chains. Alternatively, employing techniques for concatenating individual peptoid oligomers to form macromolecular species may provide a solution.

1.8.5
Develop Peptoid Shuttles for Intracellular Import of Xenobiotic Agents

Certain polycationic peptoid sequences have been shown to be capable of crossing the cell membrane (Section 1.4). The use of these "lipitoids" as transfection agents that drive the intracellular delivery of DNA and RNA [34] may be extended to other intracellular targets. The efficacy of peptides or xenobiotics that target intracellular species may be greatly enhanced upon conjugation to similar membrane-penetrating polycationic peptoids through covalent or non-covalent linkages.

1.8.6
Optimize Pharmacological Profile of Oligopeptoids

Initial studies on short peptoid oligomers have revealed relatively poor pharmacokinetic properties [18, 79]. Despite the numerous advantageous attributes of peptoids *in vitro*, there are currently no peptoid-based therapeutics. However, a more thorough exploration of peptoid sequences may reveal species with more appropriate characteristics, such as improved oral availability and slower excretion rates. Moreover, peptoid–peptide chimerae may exhibit improved behavior. As the delivery of protein therapeutics by injection has become more common, it may soon be possible to explore similar strategies for the delivery of peptoid therapeutics, mitigating some pharmacokinetic limitations. Helical peptoids for antimicrobial and lung surfactant replacement applications will also be explored for *in vivo* use, which will yield more useful information on the fate of peptoids in biological systems.

1.9
Conclusion

Efforts to investigate the questions posed here will lead to more useful peptoid designs while simultaneously leading to a better fundamental understanding of molecular recognition and sequence/structure/function relationships in non-natural, sequence-specific peptidomimetic oligomers.

1.10
References

1 PATCH, J.A. and BARRON, A.E. Mimicry of bioactive peptides by non-natural, sequence-specific peptidomimetic oligomers. *Curr. Opin. Chem. Biol.* **2002**, *6*, 872–877.

2 SIMON, R.J., KANIA, R.S., ZUCKERMANN, R.N., HUEBNER, V.D., JEWELL, D.A., BANVILLE, S., NG, S., WANG, L., ROSENBERG, S., MARLOWE, C.K., SPELLMEYER, D.C., TAN, R., FRANKEL, A.D., SANTI, D.V., COHEN, F.E., and BARTLETT, P.A. Peptoids: a modular approach to drug discovery. *Proc. Natl. Acad. Sci. USA* **1992**, *89*, 9367–9371.

3 ARMAND, P., KIRSHENBAUM, K., FALICOV, A., DUNBRACK, R.L., J., DILL, K.A., ZUCKERMANN, R.N., and COHEN, F.E.Chiral N-substituted glycines can form stable helical conformations. *Fold. Design* **1997**, *2*, 369–375.

4 KIRSHENBAUM, K., BARRON, A.E., GOLDSMITH, R.A., ARMAND, P., BRADLEY, E.K., TRUONG, K.T.V., DILL, K.A., COHEN, F.E., and ZUCKERMANN, R.N. Sequence-specific polypeptoids: A diverse family of heteropolymers with stable secondary structure. *Proc. Natl. Acad. Sci. USA* **1998**, *95*, 4305–4308.

5 MILLER, S.M., SIMON, R.J., NG, S., ZUCKERMANN, R.N., KERR, J.M., and MOOS, W.H. Comparison of the proteolytic susceptibilities of homologous L-amino acid, D-amino acid, and N-substituted glycine peptide and peptoid oligomers. *Drug Devel. Res.* **1995**, *35*, 20–32.

6 MURPHY, J.E., UNO, T., HAMER, J.D., COHEN, F.E., DWARKI, V., and ZUCKERMANN, R.N.A. combinatorial approach to the discovery of efficient cationic peptoid reagents for gene delivery. *Proc. Natl. Acad. Sci. USA* **1998**, *95*, 1517–1522.

7 KRUIJTZER, J.A.W., HOFMEYER, L.J.F., HEERMA, W., VERSLUIS, C., and LISKAMP, R.M.J. Solid-phase syntheses of peptoids using Fmoc-protected N-substituted glycines: The synthesis of (retro)peptoids of leu-enkephalin and substance P. *Euro. J. Chem.* **1998**, *4*, 1570–1580.

8 ZUCKERMANN, R.N., KERR, J.M., KENT, S.B.H., and MOOS, W.H. Efficient meth-

od for the preparation of peptoids [oligo(N-substituted glycines)] by submonomer solid-phase synthesis. *J. Am. Chem. Soc.* **1992**, *114*, 10646–10647.

9 FIGLIOZZI, G. M., GOLDSMITH, R., NG, S. C., BANVILLE, S. C., and ZUCKERMANN, R. N. Synthesis of N-substituted glycine peptoid libraries. *Meth. Enzymol.* **1996**, *267*, 437–447.

10 UNO, T., BEAUSOLEIL, E., GOLDSMITH, R. A., LEVINE, B. H., and ZUCKERMANN, R. N. New submonomers for poly N-substituted glycines (peptoids). *Tetr. Lett.* **1999**. *40*, 1475-1478.

11 BURKOTH, T. S., FAFARMAN, A. T., CHARYCH, D. H., CONNOLLY, M. D., and ZUCKERMANN, R. N. Incorporation of unprotected heterocyclic side chains into peptoid oligomers via solid-phase submonomer synthesis. *J. Am. Chem. Soc.* **2003** *125*, 8841–8845.

12 TAYLOR, E. W., GIBOONS, J. A., and BAECKMAN, R. A. Intestinal absorption screening of mixtures from combinatorial libraries in the Caco-2 model. *Pharm. Res.* **1997**, *14*, 572–577.

13 BOEIJEN, A. and LISKAMP, R. M. J. Sequencing of peptoid peptidomimeticcs by Edman degradation. *Tet. Lett.* **1998**, *39*, 3589–3592.

14 BURKOTH, T. S., BEAUSOLEIL, E., KAUR, S., TANG, D., COHEN, F. E., and ZUCKERMANN, R. N. Toward the synthesis of artificial proteins: The discovery of an amphiphilic helical peptoid assembly. *Chem. Biol.* **2002**, *9*, 647–654.

15 HEERMA, W., BOON, J.-P. J. L., VERLUIS, C., KRUIJTZER, J. A. W., HOFMEYER, L. J. F., and LISKAMP, R. M. J. Comparing mass spectrometric characteristics of peptides and peptoids – 2. *J. Mass Spec.* **1997**, *32*, 697–704.

16 HEERMA, W., VERSLUIS, C., KRUIJTZER, C. G. D., ZIGROVIC, I., and LISKAMP, R. M. J. Comparing mass spectrometric characteristics of peptides and peptoids. *Comm. Mass Spec.* **1996**, *10*, 459–464.

17 ZUCKERMANN, R. N., MARTIN, E. J., SPELLMEYER, D. C., STAUBER, G. B., SHOEMAKER, K. R., KERR, J. M., FIGLIOZZI, G. M., GOFF, D. A., SIANI, M. A., SIMON, R. J., BANVILLE, S. C., BROWN, E. G., WANG, L., RICHTER, L. S., and MOOS, W. H. Discovery of nanomolar ligands for 7-transmembrane G-protein-coupled receptors from a diverse N-(substituted)glycine peptoid library. *J. Med. Chem.* **1994**, *37*, 2678–2685.

18 GIBBONS, J. A., HANCOCK, A. A., VITT, C. R., KNEPPER, S., BUCKNER, S. A., BURNE, M. E., MILICIC, I., JAMES F. KERWIN, J., RICHTER, L. S., TAYLOR, E. W., SPEAR, K. L., ZUCKERMANN, R. N., SPELLMEYER, D. C., BRAECKMAN, R. A., and MOOS, W. H. Pharmacologic characterization of CHIR 2279, an N-substituted glycine peptoid with high-affinity binding for α1-adrenoceptors. *J. Pharm. Exp. Ther.* **1996**, *277*, 885–899.

19 NG, S., GOODSON, B., EHRHARDT, A., MOOS, W. H., SIANI, M., and WINTER, J. Combinatorial discovery process yields antimicrobial peptoids. *Bioorg. Med. Chem.* **1999**, *7*, 1781–1785.

20 KLESCHICK, W. A., DAVIS, L. N., DICK, M. R., GARLICH, J. R., MARTIN, E. J., ORR, N., NG, S. C., PERNICH, D. J., UNGER, S. H., WATSON, G. B., and ZUCKERMANN, R. N. The application of combinatorial chemistry in agrochemical discovery. In *ACS Symp. Ser. N. 774.* **2001**, American Chemical Society.

21 GARCIA-MARTINEZ, C., HUMET, M., PLANELLS-CASES, R., GOMIS, A., CAPRINI, M., VIANA, F., PENA, E. D., SANCHEZ-BAEZA, F., CARBONELL, T., FELIPE, C. D., PEREZ-PAYA, E., BELMONTE, C., MESSEGUER, A., and FERRE-MONTIEL, A. Attenuation of thermal nociception and hyperalgesia by VR1 blockers. *Proc. Natl. Acad. Sci. USA* **2002**, *99*, 2374–2379.

22 HEIZMANN, G., HILDEBRAND, P., TANNER, H., KETTERER, S., PANSKY, A., FROIDEVAUX, S., BEGLINGER, C., and EBERLE, A. N. A combinatorial peptoid library for the identification of novel MSH and GRP/Bombesin receptor ligands. *J. Receptor Signal Transduct. Res.* **1999**, *19*, 449–466.

23 THOMPSON, D. A., CHAI, B.-X., ROOD, H. L. E., SIANI, M. A., DOUGLAS, N. R., GRANTZ, I., and MILLHAUSER, G. L. Peptoid mimetics of agouti related protein. *Bioorg. Med. Chem. Lett.* **2003**, *13*, 1409–1413.

24 HAMY, F., FELDER, E. R., HEIZMANN, G., LAZDINS, J., ABOUL-ELA, F., VARANI, G., KARN, J., and KLIMKAIT, T. An inhibitor

of the Tat/TAR RNA interaction that effectively suppresses HIV-1 replication. *Proc. Natl. Acad. Sci. USA* **1997**, *94*, 3548–3553.

25 KESAVAN, V., TAMILARASU, N., CAO, H., and RANA, T. M. A new class of RNA-binding oligomers: Peptoid amide and ester analogues. *Bioconj. Chem.* **2002**, *13*, 1171–1175.

26 RYSER, H. J. P. A membrane effect of basic polymers dependent on molecular size. *Nature* **1967**. *215*, 934–936.

27 EMI, N., KIDOAKI, S., YOSHIKAWA, K., and SAITO, H. Gene transfer mediated by polyarginine requires a formation of big carrier–complex of DNA aggregate. *Biochem. Biophys. Res. Commun.* **1997**, *231*, 421–424.

28 FRANKEL, A. D. and PABO, C. O. Cellular uptake of the Tat protein from human immunodeficiency virus. *Cell* **1988**, *55*, 1189–1193.

29 VIVÈS, E., BRODIN, P., and LEBLEU, B. A truncated HIV-1 Tat protein basic domain rapidly translocates through the plasma membrane and accumulates in the cell nucleus. *J. Biol. Chem.* **1997**, *272*, 16010–16017.

30 KLIMKAIT, T., FELDER, E. R., ALBRECHT, G., and HAMY, F. Rational optimization of a HIV-1 Tat Inhibitor: Rapid progress on combinatorial lead structures. *Biotechnol. Bioeng.* **1999**, *61*, 155–164.

31 DAELEMANS, D., SCHOLS, D., WITVROUW, M., RANNECOUQUE, C., HATSE, S., DOOREN, S. V., HAMY, F., KLIMKAIT, T., CLERCQ, E. D., and VANDAMME, A.-M. A second target for the peptoid Tat/transactivation response element inhibitor CGP64222: Inhibition of human immunodeficiency virus replication by blocking CXC-chemokine receptor 4-mediated virus entry. *Mol. Pharmacol.* **2000**, *57*, 116–124.

32 WENDER, P. A., MITCHELL, D. J., PELKEY, E. T., STEINMAN, L., and ROTHBARD, J. B. The design, synthesis, and evaluation of molecules that enable or enhance cellular uptake: Peptoid molecular transporters. *Proc. Natl. Acad. Sci. USA* **2000**, *97*, 13003–13008.

33 DAVIS, M. E., Non-viral gene delivery systems. *Curr. Opin. Biotechnol.* **2002**, *13*, 128–131.

34 HUANG, C.-Y., UNO, T., MURPHY, J., LEE, S., HAMER, J., ESCOBEDO, J., COHEN, F., RADHAKRISHNAN, R., DWARKI, V., and ZUCKERMANN, R. N. Lipitoids – novel cationic lipids for cellular delivery of plasmid DNA *in vitro*. *Chem. Biol.* **1998**, *5*, 345–354.

35 NGUYEN, J. T., TURCK, C. W., COHEN, F. E., ZUCKERMANN, R. N., and LIM, W. A. Exploiting the basis of proline recognition by SH3 and WW domains: Designs of *N*-substituted inhibitors. *Science* **1998**, *282*, 2088–2092.

36 NGUYEN, J. T., PORTER, M., AMOUI, M., MILLER, W. T., ZUCKERMANN, R. N., and LIM, W. A. Improving SH3 domain ligand selectivity using a non-natural scaffold. *Chem. Biol.* **2000**, *7*, 463–473.

37 MATTERN, R. H., MOORE, S. B., TRAN, T. A., RUETER, J. K., and GOODMAN, M. Synthesis, biological activities, and conformational studies of somatostatin analogues. *Tetrahedron* **2000**, *56*, 9819–9831.

38 RUIJTENBEEK, R., KRUIJTZER, J. A. W., WIEL, W. V. D., FISCHER, M J. E., FLUCK, M., REDEGELD, F. A. M., and LISKAMP, R. M. J. Peptoid–peptide hybrids that bind Syk SH2 domains involved in signal transduction. *ChemBioChem* **2001**, *2*, 171–179.

39 HAAN, E. C. d., WAUBEN, M. H. M., GROSFELD-STULEMEYER, M. C., KRUIJTZER, J. A. W., LISKAMP, R. M. J., and MORET, E. E. Major histocompatibility complex class II binding characteristics of peptoid–peptide hybrids. *Bioorg. Med. Chem.* **2002**, *10*, 1939–1945.

40 GOTTSCHLING, D., BOER, J., MARINELLI, L., VOLL, G., HAUPT, M., SCHUSTER, A., HOLZMANN, B., and KESSLER, H. Synthesis and NMR structure of peptidomimetic $\alpha 4\beta 7$-integrin antagonists. *ChemBioChem* **2002**, *3*, 575–578.

41 GELLMAN, S. H. Foldamers: A manifesto. *Acc. Chem. Res.* **1998**, *31*, 173–180.

42 ARMAND, P., KIRSHENBAUM, K., GOLDSMITH, R. A., FARR-JONES, S., BARRON, A. E., TRUONG, K. T. V., DILL, K. A., MIERKE, D. F., COHEN, F. E., ZUCKERMANN, R. N., and BRADLEY, E. K. NMR determination of the major solution conformation of a peptoid pentamer with chiral

side chains. *Proc. Natl. Acad. Sci. USA* **1998**, *95*, 4309–4314.

43 Wu, C. W., Kirshenbaum, K., Sanborn, T. J., Patch, J. A., Huang, K., Dill, K. A., Zuckermann, R. N., and Barron, A. E. Structural and spectroscopic studies of peptoid oligomers with α-chiral aliphatic side chains. *J. Am. Chem. Soc.* **2003**, *125*, 13525–13530.

44 Wu, C. W., Sanborn, T. J., Zuckermann, R. N., and Barron, A. E. Peptoid oligomers with alpha-chiral, aromatic side chains: Effects of chain length on secondary structure. *J. Am. Chem. Soc.* **2001**, *123*, 2958–2963.

45 Wu, C., Sanborn, T., Huang, K., Zuckermann, R., and Barron, A. Peptoid oligomers with alpha-chiral, aromatic side chains: Sequence requirements for the formation of stable peptoid helices. *J. Am. Chem. Soc.* **2001**, *123*, 6778–6784.

46 Patch, J. A., Wu, C. W., Sanborn, T. J., Huang, K., Zuckermann, R. N., Barron, A. E., and Radhakrishnan, I. A peptoid oligomer with a unique, cyclic folded structure and a solvent-depend conformational switch. (Manuscript in preparation).

47 Sanborn, T., Wu, C., Zuckermann, R., and Barron, A. Extreme stability of helices formed by water-soluble poly-N-substituted glycines (polypeptoids) with alpha-chiral side chains. *Biopolymers* **2002**, *63*, 12–20.

48 Walgers, R., Le, T., and Cammers-Goodwin, A. An indirect chaotropic mechanism for the stabilization of helix conformation of a peptide in aqueous trifluoroethanol and hexafluoro-2-propanol. *J. Am. Chem. Soc.* **1998**, *120*, 5073–5079.

49 Beausoleil, E., Truong, K. T., Kirshenbaum, K., and Zuckermann, R. N. Influence of monomer structural elements in hydrophilic peptoids. In *Innovations and Perspectives in Solid Phase Synthesis and Combinatorial Libraries: Peptides, Proteins, and Nucleic Acids*, R. Epton (Ed.), **2001**, Mayflower Scientific Press: Kingswinford, UK, pp. 239–242.

50 Raguse, T. L., Lai, J. R., LePlae, P. R., and Gellman, S. H. Toward beta-peptide tertiary structure: Self-association of an amphiphilic 14-helix in aqueous solution. *Org. Lett.* **2001**, *3*, 3963–3966.

51 Zasloff, M. Magainins, a class of antimicrobial peptides from *Xenopus* skin: Isolation, characterization of two active forms, and partial cDNA sequence of a precursor. *Proc. Natl. Acad. Sci. USA* **1987**, *84*, 5449–5453.

52 Zasloff, M., Martin, B., and Chen, H.-C. Antimicrobial activity of synthetic magainin peptides and several analogues. *Proc. Natl. Acad. Sci. USA* **1988**, *85*, 910–913.

53 Porter, E. A., Wang, X., Lee, H.-S., Weisblum, B., and Gellman, S. H. Nonhaemolytic beta-amino-acid oligomers. *Nature* **2000**, *404*, 565.

54 Bechinger, B. Structure and functions of channel-forming peptides: Magainins, cecropins, melittin, and alamethicin. *J. Membr. Biol.* **1997**, *156*, 197–211.

55 Patch, J. A. and Barron, A. E. Helical peptoid mimics of magainin-2 amide. *J. Am. Chem. Soc.* **2003**, *125*, 12092–12093.

56 Robertson, B., Johansson, J., and Curstedt, T. Synthetic surfactants to treat neonatal lung disease. *Mol. Med. Today* **2000**, *6*, 119–124.

57 Goerke, J. and Clements, J. A. Alveolar surface tension and lung surfactant. In *Handbook of Physiology. Sec. 3, The Respiratory System. Vol. 3, Mechanics of Breathing: Part I*, A. Fishman, P. Macklem, J. Mead, and S. Geiger (Eds), **1986**, American Physiological Society: Bethesda, MD, pp. 247–262.

58 Wu, C. W. and Barron, A. E. Biomimetic lung surfactant replacements. In *Biomimetic Materials and Design: Interactive Biointerfacial Strategies, Tissue Engineering, and Drug Delivery*, A. K. Dillow, and A. Lowman (Eds), **2002**, Marcel-Dekker Publishers: New York, pp. 565–633.

59 Rouhi, A. M. A peptoid promise: Synthetic lung surfactant mimics may widen access to treatment of respiratory disease. *Chem. Eng. News* **2001**, *79*, 50–51.

60 Creuwels, L. A. J. M., Boer, E. H., Demel, R. A., Golde, L. M. G. v., and Haagsman, H. P. Neutralization of the positive charges of surfactant protein C. *J. Biol. Chem.* **1995**, *270*, 16225–16229.

61 JOHANSSON, J., SZYPERSKI, T., CURSTEDT, T., and WUTHRICH, K. The NMR structure of the pulmonary surfactant-associated polypeptide SP-C in an apolar solvent contains a valyl-rich alpha-helix. *Biochemistry* **1994**, *33*, 6015–6023.

62 GERICKE, A., FLACH, C. R., and MENDELSOHN, R. Structure and orientation of lung surfactant SP-C and L-alpha-dipalmitoylphosphatidylcholine in aqueous monolayers. *Biophys. J.* **1997**, *73*, 492–499.

63 BRUNI, R., TAEUSCH, H. W., and WARING, A. J. Surfactant protein B: Lipid interactions of synthetic peptides representing the amino-terminal amphipathic domain. *Proc. Natl. Acad. Sci. USA* **1991**, *88*, 7451–7455.

64 LIPP, M. M., LEE, K. Y. C., WARING, A., and ZASADZINSKI, J. A. Fluorescence, polarized fluorescence, and Brewster angle microscopy of palmitic acid and lung surfactant protein B monolayers. *Biophys. J.* **1997**, *72*, 2783–2804.

65 GORDON, L. M., HORVATH, S., LONGO, M. L., ZASADZINSKI, J. A. N., TAEUSCH, H. W., FAULL, K., LEUNG, C., and WARING, A. J. Conformation and molecular topography of the N-terminal segment of surfactant protein B in structure-promoting environments. *Protein Sci.* **1996**, *5*, 1662–1675.

66 GORDON, L. M., LEE, K. Y. C., LIPP, M. M., ZASADZINSKI, J. A., WALTHER, F. J., SHERMAN, M. A., and WARING, A. J. Conformational mapping of the N-terminal segment of surfactant protein B in lipid using C-13-enhanced Fourier transform infrared spectroscopy. *J. Peptide Res.* **2000**, *55*, 330–347.

67 SEURYNCK, S. L., JOHNSON, M., and BARRON, A. E. Simple helical peptoid analogues of lung surfactant protein B. **2003**. (Manuscript in preparation).

68 WU, C. W., SEURYNCK, S. L., LEE, K. Y. C., and BARRON, A. E. Helical peptoid mimics of lung surfactant protein C. *Chem. Biol.* **2003**, *10*, 1057–1063.

69 JENKINS, C. L. and RAINES, R. T. Insights on the conformational stability of collagen. *Nat. Prod. Rep.* **2002**, *19*, 49–59.

70 GOODMAN, M., FENG, Y., MELACINI, G., and TAULANE, J. P. A template-induced *incipient* collagen-like triple-helical structure. *J. Am. Chem. Soc.* **1996**, *118*, 5156–5157.

71 MELACINI, G., FENG, Y., and GOODMAN, M. Collagen-based structures containing the peptoid residue *N*-isobutylglycine (Nleu). 6. Conformational analysis of Gly-Pro-Nleu sequences by ^{1}H NMR, CD, and molecular modeling. *J. Am. Chem. Soc.* **1996**, *118*, 10725–10732.

72 GOODMAN, M., MELACINI, G., and FENG, Y. Collagen-like triple helices incorporating peptoid residues. *J. Am. Chem. Soc.* **1996**, *118*, 10928–10929.

73 FENG, Y., MELACINI, G., and GOODMAN, M. Collagen-based structures containing the peptoid residue *N*-isobutylglycine (Nleu): Synthesis and biophysical studies of Gly-Nleu-Pro sequences by circular dichroism. *Biochemistry* **1997**, *36*, 8716–8724.

74 MELACINI, G., FENG, Y., and GOODMAN, M. Collagen-based structures containing the peptoid residue *N*-isobutylglycine (Nleu): Conformation analysis of Gly-Nleu-Pro sequences by ^{1}H-NMR and molecular modeling. *Biochemistry* **1997**, *36*, 8725–8732.

75 JEFFERSON, E. A., LOCARDI, E., and GOODMAN, M. Incorporation of achiral peptoid-based trimeric sequences into collagen mimetics. *J. Am. Chem. Soc.* **1998**, *120*, 7420–7428.

76 KWAK, J., JEFFERSON, E. A., BHUMRALKAR, M., and GOODMAN, M. Triple helical stabilities of guest–host collagen mimetic structures. *Bioorg. Med. Chem.* **1999**, *7*, 153–160.

77 JOHNSON, G., JENKINS, M., McLEAN, K. M., GRIESSER, H. J., KWAK, J., GOODMAN, M., and STEELE, J. G. Peptoid-containing collagen mimetics with cell binding activity. *J. Biomed. Mater. Res.* **2000**, *51*, 612–624.

78 HILL, D. J., MIO, M. J., PRINCE, R. B., HUGHES, T. S., and MOORE, J. S. A field guide to foldamers. *Chem. Rev.* **2001**, *101*, 3893–4011.

79 WANG, Y., LIN, H., TULLMAN, R., CHARLES F. JEWELL, J., WEETALL, M. L., and TSE, F. T. S. Absorption and disposition of a tripeptoid and a tetrapeptide in the rat. *Biopharm. Drug. Dispos.* **1999**, *20*, 69–75.

2

β-Peptides, γ-Peptides and Isosteric Backbones: New Scaffolds with Controlled Shapes for Mimicking Protein Secondary Structure Elements

Gilles Guichard

2.1
Introduction

The remarkable and highly diverse biological activities exhibited by proteins rely on the unique capacity of these intrinsically flexible chains to fold into well-ordered and compact structures. Detailed information about these molecular and supramolecular structures is a prerequisite for the comprehension of the biological events in the living cell and this was recognized early on by Linus Pauling more than half a century ago. The formation of protein tertiary and quaternary structures relies only on a small set of distinct secondary structural elements: (i) sheets, (ii) helices, and (iii) turns. In recent years, the understanding of the folding and self-assembly processes at work in proteins, which are essentially governed by non-covalent forces, have led to major advances in the *de novo* design of individual protein secondary structure elements and protein folds from α-polypeptides [1–5].

Parallel to this active field of protein design, chemists have been creating new synthetic oligomeric materials that can self-organize spontaneously but in a controlled fashion to form defined secondary structures. Intra- and intermolecular self-organization in such oligomeric designs may result from a variety of non-covalent forces including electrostatic interactions, H-bonds, aromatic–aromatic interactions, coordination to metal ions, steric interactions or solvophobic effects. These simplified artificial systems based on short chain synthetic oligomers designed to fold into regular conformations provide useful models to study the factors that govern the formation of three-dimentional structures in biopolymers. The term "foldamer" was first proposed by Gellman to refer to such folding oligomers [6, 7]. The advances accomplished in this emerging interdisciplinary field have recently been covered in depth by Moore and coworkers [8]. According to the authors, a foldamer is "…any oligomer that folds into a conformationally-ordered state in solution, the structures of which are stabilized by a collection of non-covalent interactions between non-adjacent monomer units. There are two major classes of foldamers: single-stranded foldamers that only fold (peptidomimetics and their abiotic analogs) and multiple-stranded foldamers that both associate and fold (nucleotidomimetics and their abiotic analogs)".

Pseudo-peptides in Drug Discovery. Edited by Peter E. Nielsen
Copyright © 2004 Wiley-VCH Verlag GmbH & Co. KGaA, Weinheim
ISBN: 3-527-30633-1

In the field of peptidomimetics, seminal work by the groups of Seebach, Gellman and Hanessian has demonstrated that properties such as folding and structural diversity are not restricted to natural linear α-polypeptides but are shared by synthetic peptides consisting exclusively of higher ω-amino acids such as β- and γ-amino-acids. Short chain β- and γ-peptides with defined substitution patterns can form stable helices, sheets and turns in solution and in the solid state. Even higher ω-amino acids such as conformationally restricted δ-amino acids (e.g. sugar amino acids) have since been reported for the construction of folding oligomers. Further transformation of the β- and γ-peptide backbones by insertion of isosteric modifications results in non-peptide oligomers that in turn might well form stable and novel secondary structures. Altogether, such unnatural oligomeric scaffolds designed to reproduce or mimic essential protein structural elements, while retaining the chemical diversity of amino acid side chains could be of considerable value in the drug discovery process [9]. This chapter will cover recent developments in the field of linear peptidomimetic oligomers (building blocks synthesis, structure and function), with particular focus on the β- and the γ-peptide lineages. The synthesis, self-assembly and biological properties of related macrocyclic oligomers will not be covered in this monograph.

2.2
Molecular Organization in β-Peptide Oligomers

2.2.1
Historical Background

In 1996, research groups led by Gellman and Seebach, independently reported that enantiopure short chain β-peptides with an appropriate substitution pattern (oligomer of β-substituted β-amino acids **1** [10] and oligomer of *trans*-2-amino-cyclohexane carboxylic acid (*trans*-ACHC) **2** [6], respectively) could fold in a predict-

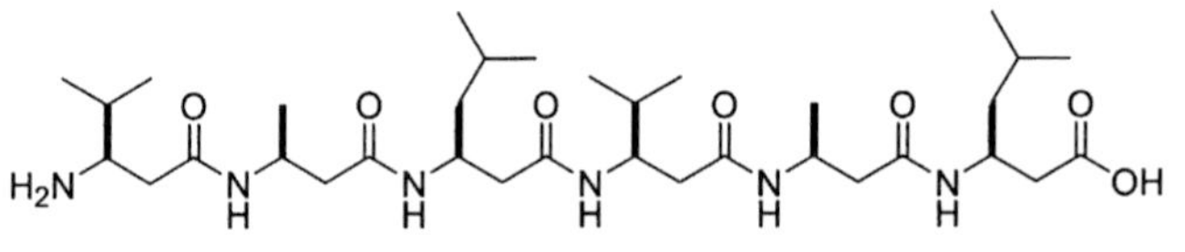

1

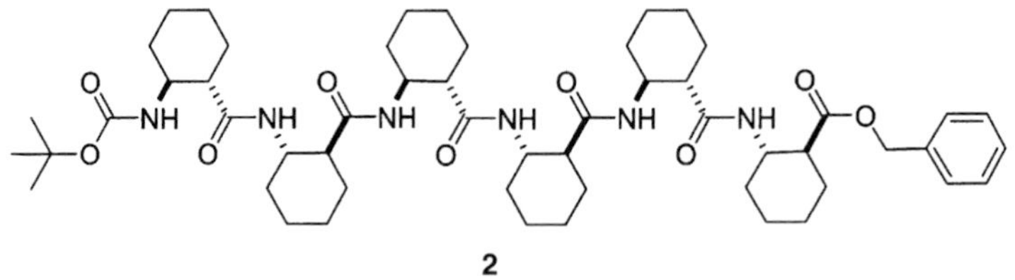

2

able manner into a highly stable and new type of helical secondary structure reminiscent of the α-helix. One year later, a flurry of commentaries highlighted the significance of the work in the fields of peptidomimetic chemistry and molecular design [11–13]. At first glance however, the result was a surprise. In the case of oligomers of β-substituted-β-amino acids, one might have rather expected an increased conformational flexibility compared to α-peptides because of possible free-rotation around the C(α)-C(β) single bond.

Before analyzing in detail the conformational behaviour of β-peptides, it is instructive to look back into the origins and the context of this discovery. The possibility that a peptide chain consisting exclusively of β-amino acid residues may adopt a defined secondary structure was raised in a long series of studies which began some 40 years ago, on β-amino acid homopolymers (nylon-3 type polymers), such as poly(β-alanine) **3** [14, 15], poly(β-aminobutanoic acid) **4** [16–18], poly(α-dialkyl-β-aminopropanoic acid) **5** [19], poly(β-L-aspartic acid) **6** [20, 21], and poly-(α-alkyl-β-L-aspartate) **7** [22–36] (Fig. 2.1).

From studies in solution and in the solid state, a number of structures both sheet-like [14] and helical [19, 21, 24–36] have been proposed over the years for these polymers. Poly(β-alanine) **3** for example crystallizes as extended chains [14] (it is reported however to be disordered in solution [15]) while poly(α-isobutyl-L-aspartate) (**7**, R=iBu) fibers have been found to form helical secondary structures [23–27]. However, conformational investigations of these polymers proved to be difficult. The case of poly(α-isobutyl-L-aspartate) is particularly representative of the difficulty in proposing correct and definitive models of secondary structure on the basis of experimental data. In initial studies, Yuki and coworkers suggested that this polymer exists as a helix in solution [22]. However, in later work using TFE solutions and solid films cast from TFE, the authors finally concluded that the structure was that of a β-sheet [23]. In 1984, X-ray diffraction of fibers cast from solution of chloroform revealed a hexagonal crystal form that was interpreted as a helix [24]. The model built from these data was a left-handed helix with 3.25 residues per turn stabilized by H-bonds closing 16-membered pseudo-rings ((M)-3.25$_{16}$ helix). In fact, three other models turn out to be compatible with

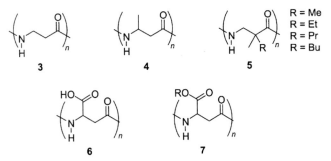

Fig. 2.1 Examples of β-amino acid homopolymers (nylon-3 type polymers) for which conformational studies have been reported

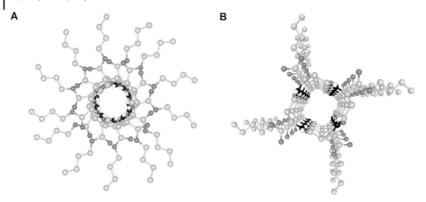

Fig. 2.2 Most favored helical structures proposed for two crystal forms of poly(α-n-butyl-β-L-aspartate) (**7**, R=Bu) [28]. (**A**) Model of a (P)-3.25_{14}-helix. (**B**) Model of a (P)-4_{18}-helix

the hexagonal crystal form. A few years later, the structure was further refined and finally reinterpreted as a right handed 3.25-helix characterized by the presence of 14-membered hydrogen-bonded pseudocycles between NH_i and $C=O_{i+2}$ ((P)-3.25_{14} helix) [25–27]. In the meantime a tetragonal form was reported for samples prepared by precipitation of the chloroform solution with ethanol that was consistent with a 4-helical conformation [25]. Again, four different models differing in their H-bonding scheme and helicity have been proposed. From further refinement, the most favored was found to be a (P)-4_{18} helix characterized by the presence of H-bonds between NH_i and $C=O_{i+3}$ [27]. These studies have since been extended to other poly(α-alkyl-β-L-aspartates) with various side chains and results revealed that the (P)-3.25_{14} and (P)-4_{18} helices are formed most frequently, the latter form being restricted to members bearing short and medium alkyl side chains (<five carbon atoms) [29–35]. These two helical structures are represented in the case of poly(α-n-butyl-β-L-aspartate) (**7**, R=Bu) in Fig. 2.2 [28].

Interestingly, it was in a different context that both Seebach and Gellman approached the field of β-peptides. Seebach's initial interest in β-peptides stemmed from their resemblance to poly(β-hydroxy alkanoates) (PHA), an ubiquitous class of biopolymers of which poly[(R)-3-hydroxybutanoic acid] (**8**, PHB) is the most common (for reviews see [37, 38]).

High molecular weight PHB (>12,000 units), which was first described more than 75 years ago is a microbial storage material (intracellular carbon source, en-

a $n=1$	**e** $n=5$
b $n=2$	**f** $n=6$
c $n=3$	**g** $n=7$
d $n=4$	**h** $n=8$

8

9

ergy storage, storage of reductase equivalents). Low molecular weight PHB (100–200 units) is found in the membrane of prokaryotic and eucaryotic cells, where it is believed to play a role in ion transport. A complex formed between PHB and calcium polyphosphate (PPi) has been shown to function as a calcium-selective ion channel in *E. coli* [39, 40]. In view of these important biological properties, and with the aim of gaining further insight into the structure of the channels consisting of PHB, Seebach and coworkers have been investigating the structures and properties of oligo(3-hydroxyalkanoates) (oligo-HA) for many years [37, 38, 41–48]. Although, PHB in the crystal state is known to adopt a 2_1-helical structure (in this nomenclature N_m describes the helix symmetry. A motif repeats after rotation about an axis by $2\pi/N$, plus a translation parallel to the screw axis of m/N the repeat in that direction) [49–51], its conformational preference in solution remains largely unknown so far (for conformational studies in solution, see [52, 53]). Examination of preferred dihedral angles found along the backbone in nine crystal structures of cyclic oligomers **9** ("oligolides"), has led Seebach and coworkers to propose, in addition to the 2_1-helix, an alternative right-handed helical conformation with about three units per turn and a pitch of 6.0 Å, namely the (P)-3_1 helix [37, 43, 45] (Fig. 2.3).

All attempts to characterize these helical conformations (solid state and liquid state NMR, optical measurements) using short chain oligo-HA have been hampered by the high flexibility of the polyester chain ($\Delta G^{\ddagger}$ for rotation about the ester CO-O bond ca. 13 kcal/mol), and by the absence of stabilizing hydrogen bond interaction [46–48]. However, the formal replacement of the main chain oxygen atoms by NH groups (moving from oligo-HA to β-peptides) was expected by Seebach to result in an increased stability of the 3_1-helical conformation as a result of possible C=O$\cdots$HN hydrogen bonds and indeed, this turns out to be the case [10].

As a prelude to the identification of novel oligomer backbones that could promote the formation of folded conformations, Gellman and his group have been exploring local structures and H-bonding in a variety of model oligoamide structures [54–63] including those derived from simple β- and γ-amino acids: β-alanine and γ-amino butyric acid derivatives, respectively [63]. Formation of stable and regular secondary structure maintained by intramolecular H-bonds (e.g. helices) requires pre-organization of the main chain (in α-polypeptides, preferred backbone conformations derive in part from minimization of Newman and Pitzer strain, as well as pseudo-allylic A(1,3) strain) to position sequentially remote H-bond donors and acceptors in close spatial vicinity so that optimal H-bonding can occur without significant conformational alteration [64]. A common feature of protein α-helices is the formation of 13-membered H-bonded rings resulting from optimal C=O$_i$$\cdots$HN$_{i+4}$ interactions. In addition, the C=O$_i$$\cdots$HN$_{i+3}$ H-bond that closes 10-membered pseudocycles is a characteristic of both the β-turn and the 3_{10}-helix. However, H-bonds between nearest neighbor amide groups such as the C5- and C7-folding patterns are not favored in polypeptides and the C7 pseudocycle is only infrequently observed in proteins of known three-dimentional structure (Fig. 2.4).

This observation led Dado and Gellman [63] to postulate that (i) such H-bonds between nearest neighbor amide groups, should they occur, would compete with

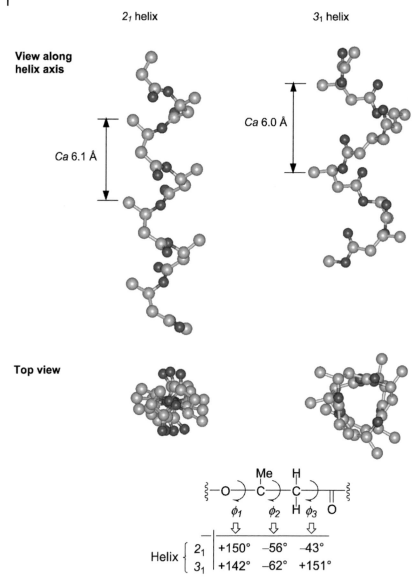

Fig. 2.3 Model of the 2_1- and 3_1-helical structures proposed for PHB chains with ideal torsion angle values. The 2_1-helix was determined by fiber X-ray diffraction of PHB [49–51] while the 3_1-helical fold was constructed by using preferred dihedral angles found along the backbone in crystal structures of cyclic oligomers **9** ("oligolides") [37, 43, 45]

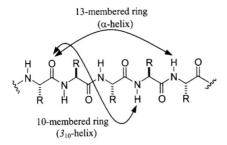

13-membered ring
(α-helix)

10-membered ring
(3_{10}-helix)

Fig. 2.4 H-bond patterns associated with helical secondary structures in α-polypeptides

long-range ordered H-bonds, and would thus be detrimental to folding and that (ii) estimation of H-bonding between nearest neighbor amide groups in simple β-alanine and γ-amino butyric acid derivatives (such as **10** and **11** respectively) could be used as a criterion to evaluate the tendency of β- and γ-peptide backbones to adopt compact and specific folding patterns.

10

11

Investigation of diamide **10** by FT-IR and NMR spectroscopy experiments revealed that no intramolecular (or intermolecular) H-bonding occurs in CH_2Cl_2 and CD_2Cl_2, respectively. Conversely, FT-IR spectra of **11** at 205 and 294 K show one NH-stretch band at 3303 and 3316 cm^{-1}, respectively, indicating the presence of an intramolecular H-bond that can close either a seven- or a nine-membered pseudocycle. Taken together, these results on model diamides tend to suggest that oligomers composed exclusively of β-amino acids rather than γ-amino acids may adopt defined secondary structures stabilized by intramolecular hydrogen bonds. However, β-alanine is perhaps not the ideal β-amino acid candidate for determining the folding propensity of β-peptides, earlier investigations in the solid state suggesting that -alanine oligomers ($n \geq 5$) would rather form β-sheet aggregates [65]. In contrast, Gellman has demonstrated in his first opus on β-peptides, that rotationally restricted 2-aminocycloalkanecarboxylic acids could provide sufficient backbone rigidity to enhance overall conformational stability [6]. In particular oligomers of *trans*-ACHC were predicted by molecular modeling to fold into a stable 3_{14} helical structure. This helical fold was experimentally assessed by X-ray diffraction and NMR analysis of synthetic β-hexapeptide **2** [6].

2.2.2
β-Amino Acids versus α-Amino Acids: An Enormous Increase in Chemical Diversity

Compared to the parent α-amino acids, β-amino acids are characterized by a much greater chemical diversity (five substitution positions versus three for α-amino acids) and conformational versatility. With the extra carbon atom in the backbone, the number of possible configurational isomers for β-amino acids increases dramatically (eight possible configurations versus two for α-amino acids) (Fig. 2.5).

A nomenclature was proposed by Seebach for the description of β-amino acids according to their substitution pattern, and for naming the resulting β-peptides [66, 67]. Enantiomerically pure β-amino acid derivatives with substituents in the 2- or 3-position are thus defined as $β^2$- and $β^3$-amino acids, respectively (abbreviated to H-$β^2$-HXaa-OH and H-$β^3$-HXaa-OH). The corresponding β-peptides built from these monomers will be named $β^2$- and $β^3$-peptides. Similarly, $β^{2,3}$-peptides consist of β-amino acid residues with substituents in both the 2- and 3-positions. Finally, peptides built from geminally disubsituted amino acids are referred to as $β^{2,2}$- and $β^{3,3}$-peptides (Fig. 2.6).

β-amino acids are not only the key monomers required for the synthesis of β-peptide foldamers, but they are also an important structural feature of many natural products including peptides and depsipeptides (for reviews, see [68, 69]) as well as of synthetic bioactive molecules and peptidomimetics (recent applications include the design of fibrinogen receptor antagonists [70], proteases inhibitors [71, 72], MHC binding peptides [73, 74], Neuropeptide Y analogs [75], DNA-binding

Fig. 2.5 β-Amino acids (five substitution positions) versus α-amino acids (three substitution positions)

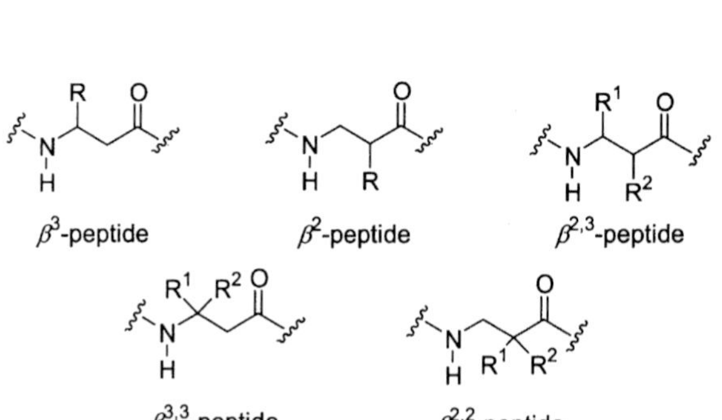

Fig. 2.6 Nomenclature proposed by Seebach to name β-peptides according to their substitution patterns [66, 67]

peptides [76] and cyclic peptide scaffolds [77]. For a general review on the incorporation of β-amino acid into peptides, see [78]). Furthermore, β-amino acids are useful and versatile precursors in the synthesis of β-lactams and β-lactam antibiotics (for review see [79]). They also serve as intermediates in the synthesis of substituted δ-lactams and δ-amino acids [80], nucleo-β-amino acid building blocks [81, 82], as well as activated carbamate derivatives for the preparation of N,N'-linked oligoureas [83]. Numerous methods have been developed for the asymmetric synthesis of β-amino acids (for reviews, see [68, 84–91]) and include the Arndt–Eistert homologation of enantiomerically pure and naturally-occurring α-amino acids, asymmetric Michael addition of chiral Li-amides to α,β-unsaturated esters, stereoselective alkylation of chiral enolates generated from β-amino esters, tetrahydropyrimidinones, amino methylation of enolates derived from chiral N-acylated oxazolidin-2-ones, enolate addition to imines and sulfinimines (Mannich-type reaction), and enantioselective hydrogenation of amino acrylates. Although a detailed account of the synthesis of β-amino acids is beyond the scope of this review, the methods currently used for the synthesis of β-amino acid building blocks en route to the preparation of β-peptides are worth being summarized.

For the synthesis and structural studies of β-peptides, Seebach and coworkers have prepared acyclic β-amino acids with different substitution pattern (including β^3-, β^2-, $\beta^{2,3}$-, $\beta^{2,2}$-, $\beta^{3,3}$- and $\beta^{2,3,3}$-amino acids) and with the side chains of natural amino acids. Gellman and his group have focused on the synthesis and use of cyclic $\beta^{2,3}$-amino acids, although in recent work they have become interested in hybrid β-peptides incorporating both cyclic and acyclic amino acids.

The Arndt–Eistert reaction (Scheme 2.1) which involves the Wolff rearrangement of diazoketones 13 (prepared from the corresponding commercially available N-protected-α-amino acids 12 by reaction of their mixed anhydrides with diazomethane; a cautionary note is warranted here: the generation and handling of diazomethane require special precautions) has been used extensively by Seebach and coworkers for the preparation of N-protected β^3-amino acids 14 and β^3-amino acid esters 15 and 16.

The mild reaction conditions permit the use of various N-protecting groups such as Boc, Z, Fmoc and are compatible with the presence of functionalized side chains (e.g. Lys, Ser, Asp, Tyr). However, partial cleavage of the Fmoc protecting group was observed under standard conditions during the homologation of N-Fmoc-protected α-amino acids. Optimized procedures for the synthesis of Fmoc-β^3-amino acids by silver-catalysed Wolff rearrangement with limited Fmoc cleavage involve the use of NMM in place of Et$_3$N [97, 98], sonication and no tertiary

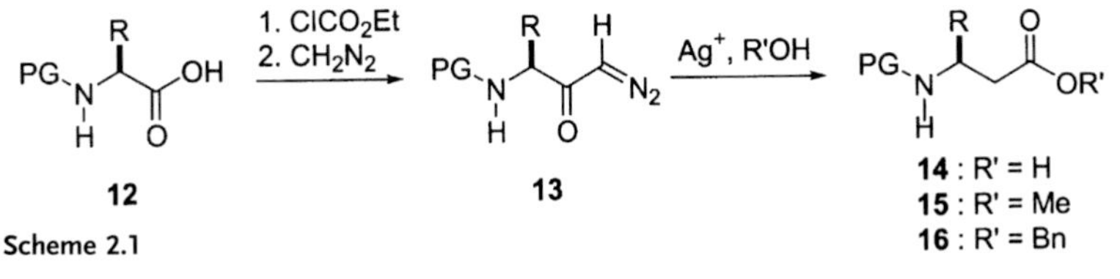

14 : R' = H
15 : R' = Me
16 : R' = Bn

Scheme 2.1

Tab. 2.1 Representative list of N-protected β^3-amino acids **14** used in the synthesis of β-peptides

Entry	PG	R	β-Amino acid 14		Yield (%)[a]	Reference
1	Boc	Me	**14a**	Boc-β^3-(S)-HAla-OH	72	92
2	Boc	iPr	**14b**	Boc-β^3-(S)-HVal-OH	81	10
4	Boc	CH$_2$OBn	**14d**	Boc-β^3-(S)-HSer(Bn)-OH	62	93
5	Boc	CH(Me)OBn	**14e**	Boc-β^3-(S)-HThr(Bn)-OH	48	94
6	Boc	CH$_2$SBn	**14f**	Boc-β^3-(S)-HCys(Bn)-OH	61[b]	95
7	Boc	(CH$_2$)$_2$SMe	**14g**	Boc-β^3-(S)-HMet-OH	59	93
8	Boc	(CH$_2$)$_2$CO$_2$Bn	**14h**	Boc-β^3-(S)-HGlu(OBn)-OH	58	96
9	Boc	(CH$_2$)$_4$NH(2-Cl-Z)	**14i**	Boc-β^3-(S)-HLys(2-Cl-Z)-OH	67	97
10	Boc	CH$_2$Ph	**14j**	Boc-β^3-(S)-HPhe-OH	37	94
11	Boc	(1H-Indol-3-yl)methyl	**14k**	Boc-β^3-(S)-HTrp-OH	55	94
12	Boc	(CH$_2$)$_2$Ph	**14l**	Boc-β^3-(S)-HHop-OH	45	97
13	Fmoc	Me	**14m**	Fmoc-β^3-(S)-HAla-OH	49	98
14	Fmoc	iPr	**14n**	Fmoc-β^3-(R)-HVal-OH	56	98
15	Fmoc	iBu	**14o**	Fmoc-β^3-(S)-HLeu-OH	57	98
16	Fmoc	CH(Me)OBn	**14p**	Fmoc-β^3-(S)-HIle-OH	60	99
17	Fmoc	CH$_2$OtBu	**14q**	Fmoc-β^3-(R)-HSer(tBu)-OH	65	98
18	Fmoc	(CH$_2$)$_2$CO$_2$tBu	**14r**	Fmoc-β^3-(S)-HGlu(OtBu)-OH	62	98
19	Fmoc	(CH$_2$)$_4$NHBoc	**14s**	Fmoc-β^3-(S)-HLys(Boc)-OH	71	98
20	Fmoc	(CH$_2$)$_3$NHBoc	**14t**	Fmoc-β^3-(S)-HOrn(Boc)-OH	38	97
21	Fmoc	CH$_2$Ph	**14u**	Fmoc-β^3-(S)-HPhe-OH	56	99
22	Fmoc	t-BuOC$_6$H$_4$CH$_2$	**14v**	Fmoc-β^3-(S)-HTyr(t-Bu)-OH	51	97
23	Fmoc	(Boc-indol-3-yl)methyl	**14w**	Fmoc-β^3-(S)-HTrp(Boc)-OH	36	99

a) Global yield from **12**.
b) Yield for the rearrangement of **13f** into **14f**.

base [100] or thermal decomposition in the absence of a tertiary base [101]. Tab. 2.1 lists the main Boc- and Fmoc-protected β^3-amino acid building blocks **14** that have been employed in the synthesis of β-peptides.

It has been shown recently that diazoketones **13** can also be alkylated to provide $\beta^{2,3}$-amino acids in moderate to high diastereoselectivities [102]. The α-methylation of Boc-protected β^3-amino acid methyl esters through the doubly lithiated derivative **17** was chosen by Seebach to gain access to both *like* and *unlike* $\beta^{2,3}$ amino acids **18** [103–105] (Scheme 2.2).

Although these Boc derivatives underwent methylation with poor selectivity (compared to 3-amino-N-benzoyl butanoates [106] and Z-protected methyl 4-phenyl-3-aminobutanoate [107]), epimers were succesfully separated by preparative HPLC or by flash chromatography. However, saponification of the methyl ester caused partial epimerization of the α-stereocenter and a two-step (epimerization free) procedure involving titanate-mediated transesterification to the corresponding benzyl esters and hydrogenation was used instead to recover the required Boc-$\beta^{2,3}$-amino acids in enantiomerically pure form [104, 105]. N-Boc-protected $\beta^{2,3}$-amino acids **19** and **20** for incorporation into water-soluble β-peptides were pre-

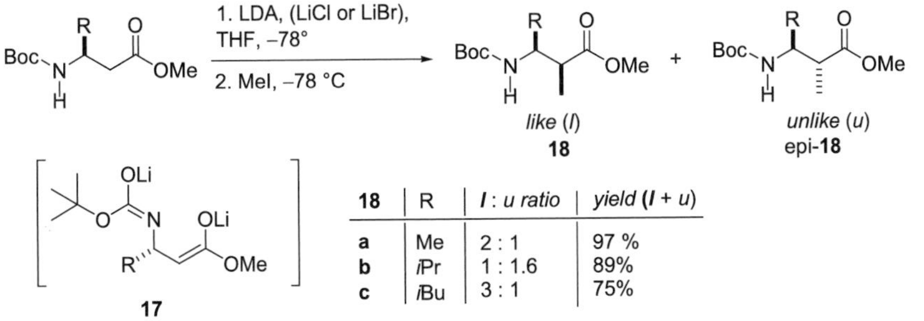

Scheme 2.2

1. LDA, (LiCl or LiBr), THF, −78°
2. MeI, −78 °C

Boc-N(H)-CH(R)-CH2-C(O)-OMe → Boc-N(H)-CH(R)-CH(CH3?)-C(O)-OMe

like (*l*)
18

unlike (*u*)
epi-**18**

17

18	R	*l* : *u* ratio	yield (*l* + *u*)
a	Me	2 : 1	97 %
b	*i*Pr	1 : 1.6	89%
c	*i*Bu	3 : 1	75%

pared in a related manner starting from Boc-β³-HSer(Bn)-OMe and Boc-β³-HCys(Bn)-OMe, respectively [108]. Other acyclic β²,³-amino acids utilized for the design of β-peptide foldamers include *anti*-compounds **21** and **22** [109] and 3-amino-2-hydroxy acid residues **23–25** [110–112].

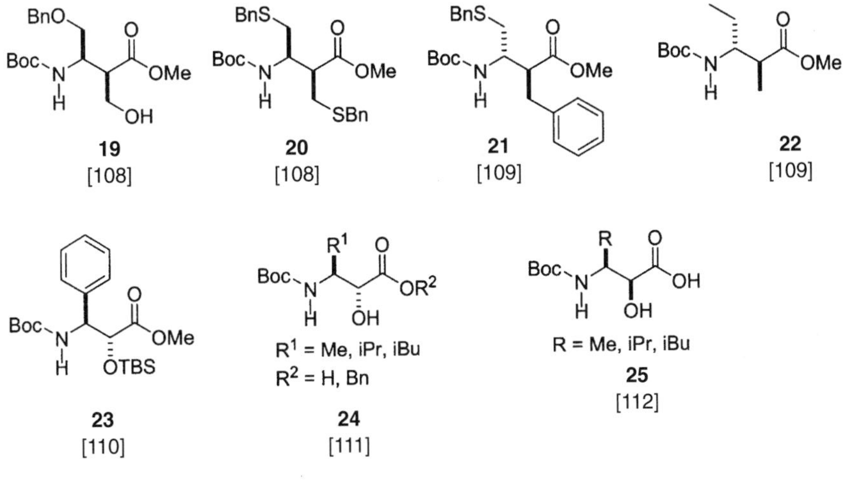

19
[108]

20
[108]

21
[109]

22
[109]

23
[110]

24
[111]
R¹ = Me, iPr, iBu
R² = H, Bn

25
[112]
R = Me, iPr, iBu

Alternatively, to avoid difficult separation of diastereomers **18a** and **epi-18a**, the Fmoc-β²,³-amino acid of *unlike* configuration **26** can be obtained as a single diastereoisomer in a three-step reaction sequence via conjugate addition of the Li-amide derived from (S)-N-benzyl-1-phenylethylamine (Davies methodology [113]) to *tert*-butyl tiglate [105] (Scheme 2.3).

The synthesis of the enantiomerically pure α-selenophenyl esters of β²,³-amino acid **27** (alkylation of the dianion of Boc-β³-HAla-OH with phenylselenium bromide gave exclusively the β²,³-amino acid of *like* configuration) and the corresponding β²,³-tri-peptide **28** and has been reported by Hanessian and coworkers [114, 115]. These derivatives undergo free radical C-allylation with allyltribuylstannane to give the corresponding *anti* type product **29** and **30** with a high stereose-

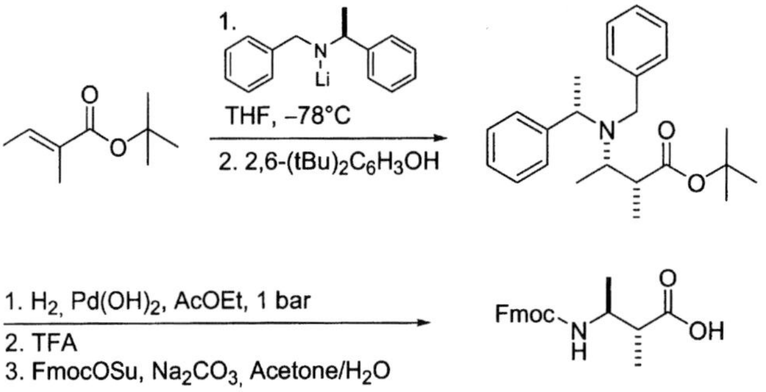

Scheme 2.3

lectivity. The remarkable level of 1,2 induction observed can be rationalized by considering intramolecular H-bonding as the main stereocontrolling factor; the crystal structure of **27** [115] reveal a C6-folding pattern conformation stabilized by an intramolecular H-bond (Scheme 2.4).

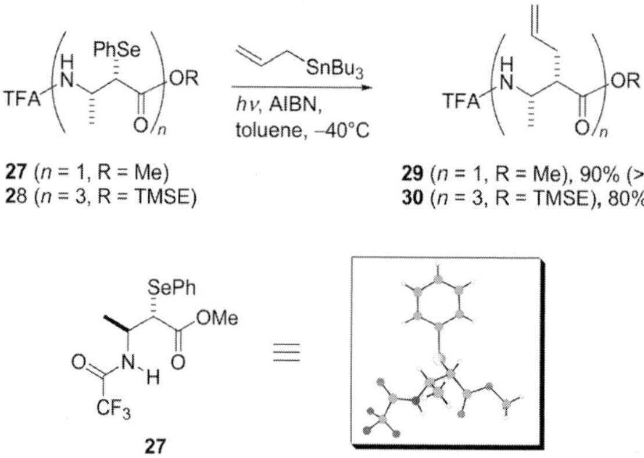

27 (*n* = 1, R = Me)
28 (*n* = 3, R = TMSE)

29 (*n* = 1, R = Me), 90% (>98:2)
30 (*n* = 3, R = TMSE), 80% (>95:5)

27

Scheme 2.4

Recent efforts in the development of efficient routes to highly substituted β-amino acids based on asymmetric Mannich reactions with enantiopure sulfinyl imine are worthy of mention. Following the pioneering work of Davis on *p*-toluenesulfinyl imines [116], Ellman and coworkers have recently developed a new and efficient approach to enantiomerically pure *N-tert*-butanesulfinyl imines and have reported their use as versatile intermediates for the asymmetric synthesis of amines [91]. Addition of titanium enolates to *tert*-butane sulfinyl aldimines and ketimines **31** proceeds in high yields and diastereoselectivities, thus providing general access to β^3-, $\beta^{3,3}$-, $\beta^{2,3}$-, $\beta^{2,3,3}$-, $\beta^{2,2,3}$-, $\beta^{2,2,3,3}$-amino acids **32** (Scheme 2.5)

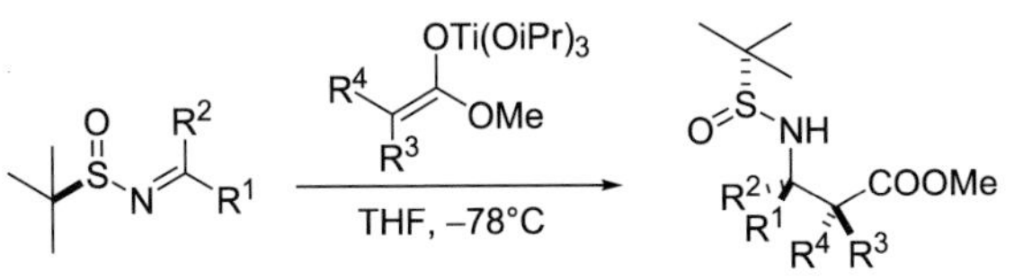

Scheme 2.5

[117, 118]. In this study, it was shown that the *N-tert*-butanesulfinyl group was not only an imine-activating and a chiral-directing group, but was also a versatile temporary amine protecting group during β-peptide synthesis.

Constrained $\beta^{2,3}$-amino acids in which the C(2)-C(3) bond is part of a small ring (5- and 6-membered rings) have been used extensively by Gellman (*trans*-2-amino-cycloalkanecarboxylic acids **33–42**) and others (**43** and **44**) for the preparation of stable β-peptide foldamers (Fig. 2.7).

The chemistry of 2-aminocycloalkanecarboxylic acids was reviewed recently [131]. Known methods to prepare ACPC and ACHC derivatives in enantiomerically pure form include resolution of racemates, biocatalyzed transformations, and enantioselective syntheses. Gellman and his groups have investigated and compared several of these methods [119–122, 124–128] and found that the one reported by Palmieri and coworkers [132] and further implemented by Xu et al. [133] for the synthesis of enantiopure *cis*-ACHC, was amenable to the preparation of enantiopure *trans*-ACPC (**33**), APC (**34**), ACHC (**40**) and APiC (**42**) derivatives in large scale for peptide synthesis. The method is general and does not necessitate chromatographic purification of the intermediates, only recrytallization procedures are employed. The approach (exemplified for the synthesis of Fmoc-APiC(Boc)-OH (**42**) in Scheme 2.6 [128]) relies on the diastereoselective reduction of β-amino esters generated from (*R*)- or (*S*)-α-methylbenzylamine (chiral auxiliary and nitrogen source) and cyclic β-ketoesters.

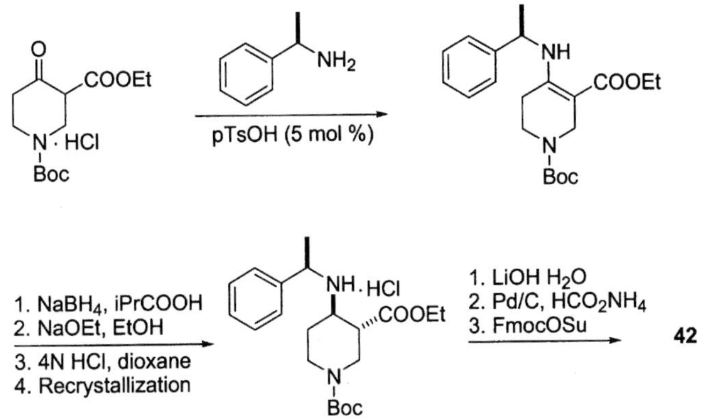

Scheme 2.6

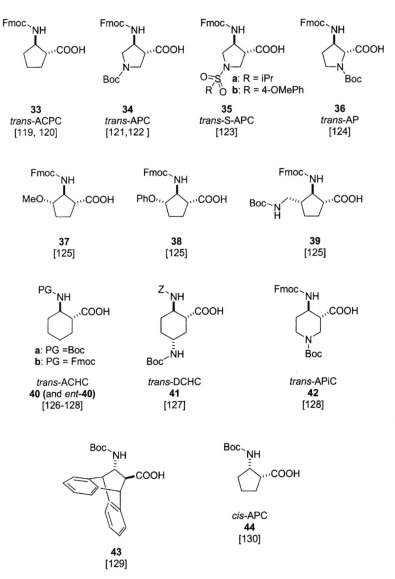

Fig. 2.7 Examples of constrained aminocycloalkane carboxylic acid derivatives for the synthesis of corresponding β-peptides

The reduction produces a mixture of four diasteomers in which a *trans-* (ACPC derivatives) or a *cis-* isomer (ACHC derivatives) is the major product. In the former case, treatment of the mixture with 4 N HCl in dioxane led to the precipitation of the *trans* isomer hydrochloride salt. One recrystallization from acetonitrile enriched the optical purity to ≥99% *de* [120, 122] In the latter case, epimerization of the mixture first, followed by the same two-stage crystallization process furnished ACHC derivatives as single *trans* isomers [128].

With the aim of preparing β^2-peptides and mixed β^2/β^3-peptides as well as allowing for direct comparison between β^2-peptides and β^3-peptides, Seebach and coworkers have prepared a number of β^2-analogs ((S) and (R)-2-(aminomethyl)alkanoic acids) **45–51** of previously reported β^3-amino acids (Fig. 2.8) [66, 97–99, 104, 134].

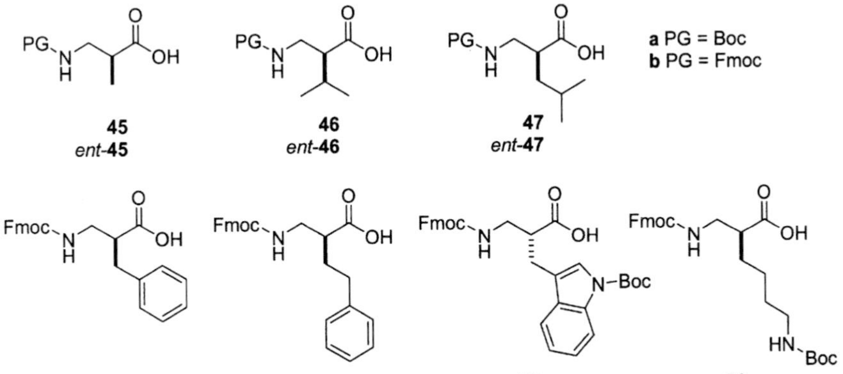

Fig. 2.8 β^2-amino acids synthesized for incorporation into β-peptides [66, 97–99, 104, 134]

These amino acids were initially synthesized by asymmetric aminomethylation of optically pure (R)- and (S)-N-Acyl-4-phenylmethyl)oxazolidin-2-ones **52** through TiCl$_4$-enolates (Evans methodology [135]) with (benzoylamino)methylchloride or benzyl N-(methoxymethyl)carbamate [66, 97–99, 104]. Hydrolytic removal of the auxiliary yielded the N-protected (benzoyl or Z) amino acid **54**. Deprotection afforded the free amino acid which was converted to the required Boc- or Fmoc-protected derivatives (Scheme 2.7).

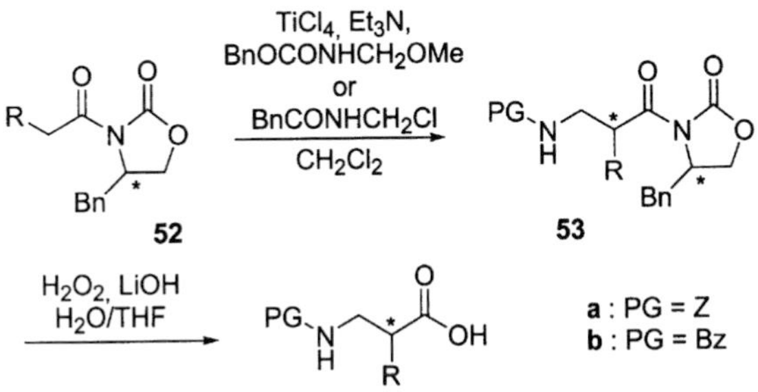

Scheme 2.7

Alternative routes to β^2-amino acids have also been explored and involve, stereo-selective alkylation of chiral derivatives of β-alanine [136–140], Curtius rearrangement of enantiomerically pure and regioselectively protected substituted-succinic acids [134, 141, 142] (the approach is also suitable for the synthesis of β^3-amino acids [143]), or the formation of chiral isoxazolidinone intermediates [144].

Both achiral and chiral $\beta^{3,3}$-amino acid (**55** and **59**), $\beta^{2,2}$-amino acid (**56–58** and **60**) as well as $\beta^{2,2,3}$-amino acids **61** have been prepared for incorporation into peptides and for conformational studies by Seebach and coworkers (for a review on the synthesis of enantiopure geminally disubstituted β-amino acids, see [89]) (Fig. 2.9). Boc-β^3-HAib-OH (**55**) was prepared by the Michael addition of NH_3 to 3-methylbut-2-enoic acid [145] and not by homologation of Boc-Aib-OH as might be expected. The reason is that both the preparation and rearrangement of diazo-ketones derived from a,a-disubstituted a-amino acids proceeds with low yield [146]. Double methylation of Boc-protected methyl 3-aminopropanoate followed by saponification of the methyl ester gave **56** in 55% overall yield [145]. A similar approach was used to prepare chiral $\beta^{2,2,3}$-amino acids **61** [147]. The synthesis of **57** ($\beta^{2,2}Ac_6c$) and **58** ($\beta^{2,2}Ac_3c$) began with the dialkylation of methyl 1-cyanoace-tate with 1,5-dibromopentane and 1,2-dibromoethane, respectively. Hydrogenation of the resulting 1-cyanocyclolalkanecarboxylates with Raney-Ni, Boc protection and saponification afforded the N-Boc-protected amino acids in 51 and 38% yield, re-spectively [145]. Enantiopure geminally substituted amino acids **59** and **60** were prepared from a common dialkylated succinate precursor via a regioselective Cur-tius rearrangement [147].

Polyproline [148, 149], proline-oligomers [150] and oligomers of N-alkylated gly-cine with chiral side chains [151, 152] adopt regular non-hydrogen bonded helical structures. By analogy, β-peptides incorporating N-methylated β-amino acid **62** [92] as well as homologs of proline **63–65** [153–155] have been synthesized with the aim of investigating the role of H-bonding in stabilization of β-peptide secondary structures (Fig. 2.10). β^3-amino acids **62** and **63** were prepared by homologation of the corresponding a-amino acids using the Arndt–Eistert reaction (*vide supra*) [92,

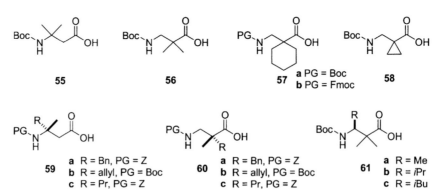

55   **56**   **57**   a PG = Boc   **58**
    b PG = Fmoc

59   a R = Bn, PG = Z   **60**   a R = Bn, PG = Z   **61**   a R = Me
    b R = allyl, PG = Boc   b R = allyl, PG = Boc   b R = iPr
    c R = Pr, PG = Z   c R = Pr, PG = Z   c R = iBu

Fig. 2.9 Geminally disubstituted amino acids synthesized to evaluate the preferential conforma-tions of $\beta^{2,2}$- and $\beta^{3,3}$-peptides

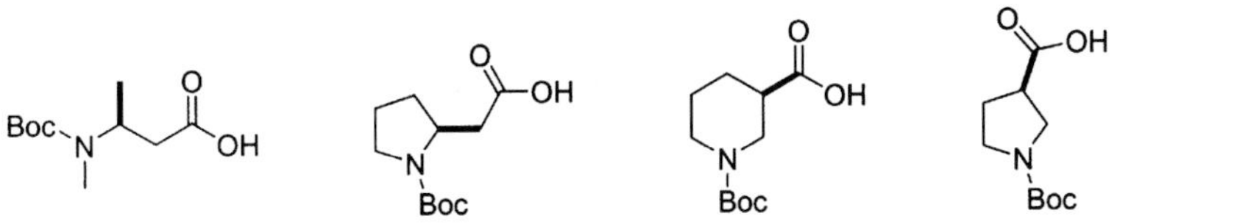

62

Boc-β³-HPro-OH
63

Boc-β²-HPro-OH
Nipecotic acid (Nip)
64

Pyrrolidine-3-carboxylic acid
(Pca)
65

Fig. 2.10 *β*-Amino acids used to investigate non-hydrogen-bonded structures

153]. Enantiopure (*R*)- and (*S*)-nipecotic acid (Nip) derivatives **64** were obtained following classical resolution of ethyl nipecotate with either enantiomer of tartaric acid and successive recrystallization of the corresponding salts [153, 154, 156] or by resolution of racemic nipecotic acid with enantiomerically pure camphorsulfonic acid [154]. *N*-Boc protected pyrrolidine-3-carboxylic acid (PCA) **65** for the synthesis of homo-oligomers [155] was prepared by Gellman from *trans*-4-hydroxy-L-proline according to a known procedure [157].

2.2.3
Helical Folds

The conformation of *α*-polypeptides can be described by three main torsion angles *ω*, *φ*, and *ψ*, the *ω* angle corresponding to the peptide bond being generally restricted to values close to 180° (*trans* geometry) or 0° (*cis* geometry). The additional torsion angle generated by the insertion of methylene units into *β*-amino acid-containing peptides is denoted $θ_1$ according to the nomenclature proposed by Balaram (Fig. 2.11) [158].

Examination of the backbone torsion angles in a number of crystal structures of *β*-alanine-containing peptides reveals that the conformation around the C(*α*)-C(*β*) bond of *β*-alanine residues is essentially *gauche* or *trans* (*anti*) with $θ_1$ values close to ±60° or 180°, respectively [158]. Populating the *gauche* conformation of *β*-ami-

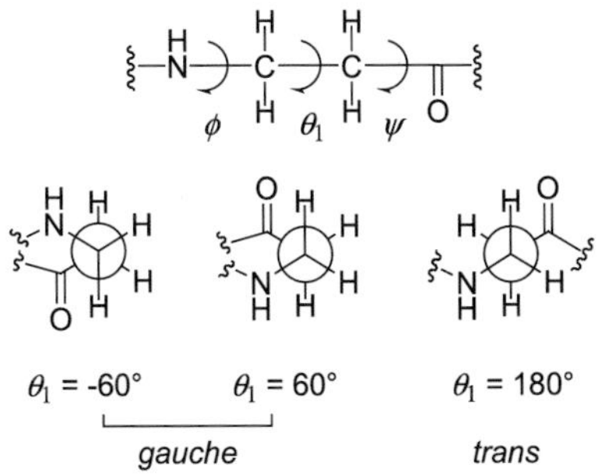

$θ_1 = -60°$ $θ_1 = 60°$ $θ_1 = 180°$

gauche   *trans*

Fig. 2.11 Preferential conformations around the C(*α*)-C(*β*) bond in the crystal structures of *β*-alanine-containing peptides [158]

no acid residues in β-peptides will ultimately lead to folded structures (helix, turn), extended structures being accessible from *trans* conformers. In-depth studies by the groups of Seebach and Gellman have delineated the substitution pattern requirements for promoting *gauche* conformation in constituent β-amino acids. Strong restriction of the conformational freedom towards *gauche* conformers in β-peptides is obtained with cyclic and acyclic $β^{2,3}$-amino acids of the *like* configuration (e.g. **18** and **33–43**, respectively) and to a lesser extent with $β^3$- and $β^2$-amino acids (e.g. **14** and **45–51** respectively). 3-Amino-2-hydroxy acids of *unlike* configuration (e.g. **24**) [111] and achiral cyclic $β^{2,2}$-amino acids (e.g. **57** and **58**) [145] are also reported to favor a gauche conformation. Thus far, four helical shapes (namely, the 3_{14} [6, 10, 103, 104, 126, 159], 2.5_{12} [119, 160], so-called 10/12- [104, 161, 162] and 2_8 helices [163, 111]) have been identified in β-peptides depending upon the substitution pattern of the constituent β-amino acid residues. This illustrates the remarkable structural diversity of these non-natural bio-oligomers.

2.2.3.1 The 3_{14}-Helix

The recognition that short chain β-peptides can form regular secondary structures initially came from detailed conformational analysis of $β^3$-peptides **1** and **66** (which incorporates a central (2S,3S)-3-amino-2-methylbutanoic acid residue) by NMR in pyridine-d_5 and CD$_3$OH [10, 103, 164] and homooligomers (as short as four residues) of *trans*-2-amino-cyclohexanecarboxylic acid (*trans*-ACHC) (e.g. hexamer **2** for the (S,S) series) by NMR and X-ray diffraction [6, 126, 159].

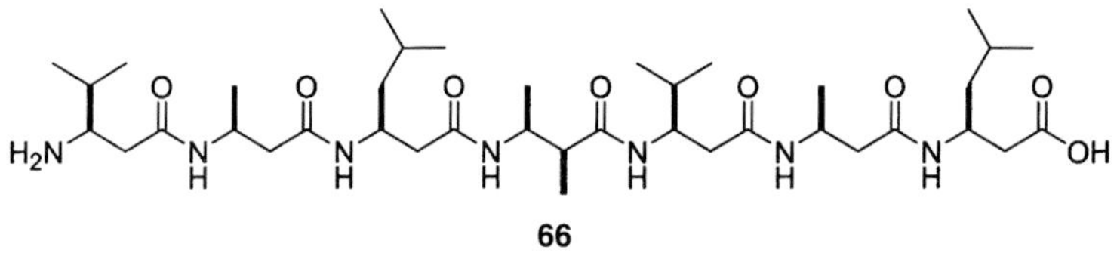

66

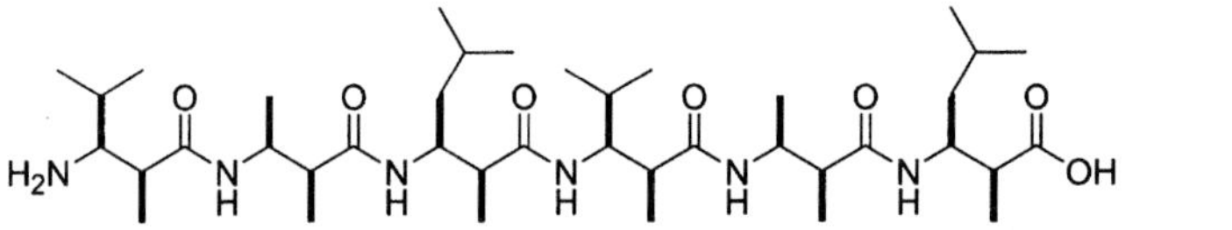

67

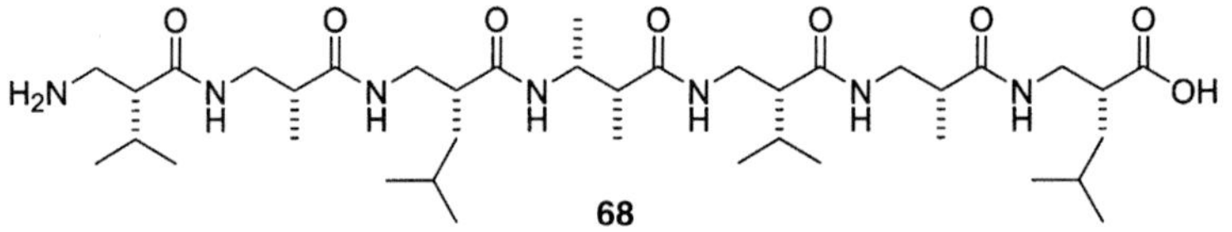

68

These studies conducted in solution and in the solid state revealed a common 3_{14}-helical fold stabilized by H-bonds closing 14-membered rings formed between NH_i and $C=O_{i+2}$ (see Fig. 2.12A and C). It is noteworthy that the 3_{14}-helix of β^3-peptides with L-amino acid-derived chirality centers (Fig. 2.12 A) and the α-helix have opposite polarity and helicity.

$\beta^{2,3}$-peptides consisting of acyclic $\beta^{2,3}$-amino acids of *like* configuration (e.g. **67**) and β^2-peptides (e.g. **68**) also adopt a 3_{14}-helical structure in solution. The helix ((M)- for **1**, **2**, **66**, and **67**; (P)- for **68**) is characterized by a rigid *synclinal* arrangement around C(α)-C(β) bonds (mean values for dihedral angles θ_1 in compounds **66** and **2** (Tab. 2.2) are 62.6° and 55.6°, respectively) with positions along the helix axis occupied exclusively by H-atoms (H^{Si} or H^{Re} protons in (P)- and (M)-helices, respectively). For steric reasons, side-chains can be tolerated only at lateral positions (Fig. 2.13). As a consequence, β-peptides consisiting of geminally disubstituted amino acids or of $\beta^{2,3}$-amino acids of *unlike* configuration will not fold into a 3_{14}-helix.

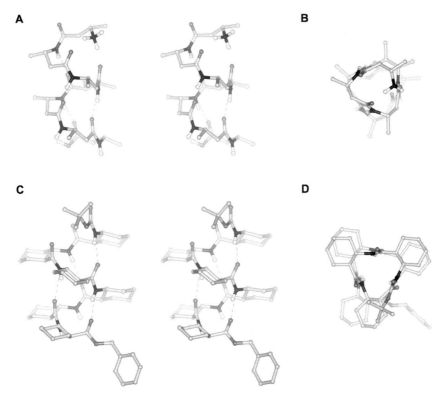

A **B** **C** **D**

Fig. 2.12 The β-peptide 3_{14}-helical structure. (**A**) Stereo-view along the helix axis of the left-handed 3_{14}-helix formed by β^3-peptide **66** in solution as determined by NMR in CD_3OH (adapted from [103, 164]). Side-chains have been omitted for clarity. (**B**) Top view.

(**C**) Stereo-view along the helix axis of the left-handed 3_{14}-helix formed by β^3-peptide **2** consisting of (S,S)-trans-2-amino-cyclohexanecarboxylic acid (trans-ACHC) residues in the solid state as determined by X-ray diffraction (adapted from [6, 126]). (**D**) Top view

Tab. 2.2 Comparison of backbone dihedral angles between 3_{14}-helical β-peptides **2** and **66**

Residue	β-Peptide 2			β-Peptide 66		
	ϕ (°)	θ_1 (°)	ψ (°)	ϕ (°)	θ_1 (°)	ψ (°)
1	−124.9	52.1	−125.3	−	67.5	−135.3
2	−140.3	52.7	−117.0	−115	57.5	−157.0
3	−156.2	55.7	−130.0	−121.7	69.2	−143.0
4	−142.0	53.6	−135.7	−127.5	52.3	−133.6
5	−131.9	57.4	−130.0	−131.8	68.1	−155.1
6	−152.3	61.9	−	−119.5	61.9	−158.4
7	−	−	−	−101.2	61.9	−

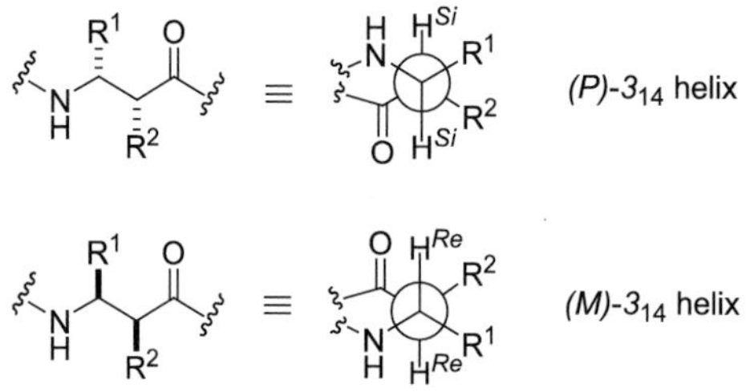

Fig. 2.13 Preferred synclinal backbone conformation around the central $C(\alpha)$-$C(\beta)$ bond in the 3_{14}-helix. For steric reasons, axial positions can be occupied only by H-atoms

Examination of the top view of the helix indicates that side-chains of residues i and $i+3$ are located almost on top of each other (Fig. 2.12 B and D) and suggest that hydrophobic interactions between overlapping aliphatic side-chains could play a significant role in stabilizing the overall structure. The distance between facing $C(\alpha)$ atoms at positions i and $i+3$ is approximately 4.8 Å. The helix is compact with a diameter of ca. 4.7 Å, slightly larger than that of the α-helix (4.3 Å).

The thermodynamic stability of the 3_{14}-helix in addition to the requirements for 3_{14}-helix formation have been studied extensively by NMR spectroscopy, circular dichroism and molecular dynamics.

Several parameters accessible by NMR are used to assess the presence of a regular 3_{14}-helical secondary structure in solution [10, 103, 104, 159]. These include : (i) *large* $^3J(NH, H–C(\beta))$ *values characteristic of a nearly antiperiplanar arrangement between NH and H–C(β)*, (ii) *chemical shift differences between diastereotopic H-C(α) protons and strong differentiation between vicinal coupling constants of each pair of diastereotopic H-C(α) protons in β³-peptides*, (in β-peptides consisting of $\beta^{2,3}$-amino acids of *like* configuration, the restricted rotation around $C(\alpha)$-$C(\beta)$ bonds is assessed by large ^3J(H-C(α), H-C(β) values indicative of a nearly antiperiplanar arrangement between H-C(α) and H-C(β) protons), and (iii) *a typical NOE pattern consisting of (i, i+2) and (i, i+3) interresidue NOE connectivities* (representative NOEs include $NH_i/H–C(\beta)_{i+2}$; $NH_i/H–C(\beta)_{i+3}$; H^{Re}-$C(\alpha)_i/H$-$C(\beta)_{i+3}$ for (M)-helix (H^{Si}-$C(\alpha)_i/H$-$C(\beta)_{i+3}$ for (P)-helix)). The rate of amide proton exchange in CD_3OD

[6, 10, 104], the variation of ^3J(NH, H-C(β)) values as a function of temperature [164] as well as the temperature coefficients ($\Delta\delta/\Delta T$) [159, 164] have been used to gain information about the conformational stability of β-peptides. For β^3- and β^2- hexa- and heptapeptides, the half-lives of the most protected protons (central residues) in H/D exchange experiments are typically between 1.5 and 2 h [10, 104]. In contrast, half-lives of amide proton exchange of up to several days have been reported for $\beta^{2,3}$ hexapeptides **2** and **67** [104, 159] thus suggesting an increased stability of the 3_{14}-helix compared to that of the corresponding β^2- and β^3-peptides. The ^3J(NH, H-C(β)) values extracted from ^{1}H-NMR spectra of **66** recorded in CD$_3$OH between 298 and 353 K with 10-K increments, show little temperature dependence, and suggest that the overall 3_{14}-helical structure is still present to a large extent at 353 K [164].

CD spectroscopy allows the study in solution of the extrinsic asymmetry resulting from preferential conformations of optically active molecules with chromophores and has been used extensively to gain rapid structural information on biomacromolecules (proteins, nucleic acids, polysaccharides), and α-polypeptides constituted from α-amino acids. Similarly, in the field of β-peptides, circular dichroism in combination with NMR spectroscopy has proven to be well adapted to the characterization of helical conformations in organic (MeOH, trifluoroethanol, acetonitrile) and aqueous solutions. 3_{14}-Helical β-peptides composed essentially of acyclic amino acid residues display a common CD signature with one extremum at ca. 215 nm (negative for (M)-helices and positive for (P)-helices) and the other of opposite sign at 195 nm [10, 98, 103–105]. β-Peptides consisting exclusively of cyclic $\beta^{2,3}$-peptides (e.g. **2**) essentially display a broad extremum at 215–217 nm with no zero crossing [126, 165]. This discrepancy at lower wavelengths between the two families of β-peptides is still not yet fully understood and might reflect a different equilibrium between the 3_{14}-helix and the interconverting folding patterns. From the numerous CD studies performed in MeOH, a wealth of information has been gathered on the requirements for 3_{14}-helix formation and stabilization. The effect of sequence and length as well as the type of β-amino acid constituent on the stability of 3_{14}-helices formed by β-peptides can be deduced very rapidly by comparison of the mean-residue ellipticity values near 215 nm (see Tab. 2.3):

(i) *The mean residue ellipticity near 215 nm increases with β-peptide chain length to reach an asymptotic value.* While a minimal chain length (ca. six residues) is required for helix formation in β^3-peptides (the CD spectrum of a β^3-trimer is not yet characteristic of the 3_{14}-helix [10]), significant backbone preorganization is already present in a dimer of *trans*-2-amino-cyclohexane carboxylic acid (*trans*-ACHC) [126].

(ii) *Temperature-dependent-CD measurements suggest that 3_{14}-helical β-peptides undergo non-cooperative unfolding upon heating.* The intensitiy of the extremum at 215 nm for peptide **66** decreases linearly (by ca. 12% per 20 K) and no cooperative break-up of the structure is observed between 295 and 333 K. This is in contrast to the unfolding characteristics of α-polypeptides where cooperative unfolding is often observed. The scan which is reversible suggests that

Tab. 2.3 Probing the 3_{14}-helical structure of β-peptides by circular dichroism in MeOH (identification of the 3_{14}-helix CD signature and mean-residue ellipticity (in deg cm^2 dmol^{-1}) of the extremum at $\lambda_1 =$ ca. 215 nm)

Entry	β-Peptide		3_{14}-helix CD pattern	$[\theta]_{ca.\ 215\ nm}$	Reference
1		1	✓	−9,000	10, 105
2		69	✓	−11,833	105
3		70	No	–	105
4		71	✓	−1,500	105

Tab. 2.3 (cont.)

Entry	β-Peptide		3_{14}-helix CD pattern	$[\theta]_{ca.\ 215\ nm}$	Reference
5		72	No	–	161
6		66	✓	−15,333	105
7		73	✓	−17,166 (at 209 nm)	108
8		2	✓	−20,000	126

Tab. 2.3 (cont.)

Entry	β-Peptide		3_{14}-helix CD pattern	$[\theta]_{ca.\ 215\ nm}$	Reference
9	*(structure)*	74	✓	+6,600	66
10	*(structure)*	68	✓	+18,333	104
11	*(structure)*	75	No	–	103
12	*(structure)*	76	✓	ca. –2,500	170

Tab. 2.3 (cont.)

Entry	β-Peptide		3₁₄-helix CD pattern	[θ]ca. 215 nm	Reference
13		77	✓	−7,333	105
14		78	✓	−7,500	95
15		79	✓	ca. +15,000	168
16		80	No	–	168

the unfolding and folding route of the helix must be reversible [164]. Reversible β-peptide folding was also succesfully investigated by molecular dynamics simulations [166].

(iii) *Fully unprotected β-peptides form more stable helices than the corresponding protected ones.* The CD spectra of the deprotected hexapeptide **1** and heptapeptide **69** (as TFA salts) show the typical CD signature with a minimum near 215 nm (entry 1 and 2, Tab. 2.3). However, the fully protected β^3-hexapeptide **70** does not form the helix (entry 3), a seventh residue being necessary for the signal of the helix to appear (peptide **71**, entry 4) [105]. In helical α-polypeptides, interactions between helix dipoles and charged groups close to the ends are known to play an important role in stabilizing (destabilizing) the helix [167]. In β-peptides it is likely that the positively charged N-terminus (NH_3^+) can stabilize the 3_{14}-helix through attractive electrostatic interactions with the negative dipole moment at the N-terminus of the helix.

(iv) *Most mixed -peptides (e.g. **72**, entry 5) containing both β^2- and β^3-amino acids do not fold into the 3_{14}-helix.* They present a distinct CD pattern that has been ascribed to another helical fold (see Section 2.2.3.2) [104, 161, 162].

(v) *Cyclic and acyclic $\beta^{2,3}$-amino acids of like configuration can exert a strong stabilizing effect on the helical structure* [105, 108, 168]. The mean residue ellipticity near 215 nm for peptides **66** and **73** which incorporates a central $\beta^{2,3}$-amino acid is greater than that of the corresponding β^3-peptide **69** (entry 2 versus entries 6 and 7).

(vi) *Helices formed by homooligomers of trans-2-amino-cyclohexane carboxylic acid (trans-ACHC) are exceptionally stable* as indicated by the mean ellipticity values near 215 nm of ca. 20,000 $deg\,cm^2\,dmol^{-1}$ for hexamer **2** (entry 8) [126].

(vii) *The 3_{14}-helix formed by β^2-peptides appears to be less stable than that formed by the corresponding β^3-peptides* (entry 1 versus entry 9, Tab. 2.3) [66, 169]. This difference might arise from the smaller steric hindrance between NHs and neighboring side-chains in β^3-peptide as compared to CO and neighboring substituents in β^2-peptides. However this stability difference seems to be reduced upon inserting a $\beta^{2,3}$-amino acid residue and increasing chain length (entry 10 versus entry 6) [98, 104].

(viii) *As expected, single point inversion of the configuration or introduction of geminally disubstituted amino acid residue into the β-peptide sequence is not tolerated* and results in the loss of the CD pattern associated with the 3_{14}-helical structure (compare entry 2 and entry 11) [103].

(ix) *Weakening the H-bond pattern or removing i/i+3 side-chain–side-chain interactions causes destabilization of the 3_{14}-helix.* Although sterically compatible with the helix, the replacement of the central residue in a β^3-heptapeptide by an hydroxy acid residue (NH→O, **76**) dramatically reduces the mean residue ellipticity values near 215 nm. The effect is less pronounced when the modification is incorporated at the 2 or 6 position [170]. Albeit tolerated, the substitution of a 3-aminopropanoyl unit (missing side-chain, **77**) for the central amino acid residue in **2** is destabilizing (entry 2 versus entry 13) [164].

(x)   The 3_{14}-helical structure of β^3-peptides is maintained upon covalent disulfide bridge formation between side chains at i and i+3 positions. In MeOH, the conformationally constrained β-hexapeptide **78** (entry 14) with a disulfide bridge between positions 2 and 5 displays a CD pattern similar to the typical pattern of the 3_{14}-helix, slightly blue shifted compared to that of the corresponding linear precursor [95, 171].

(xi) β-Amino acids with side-chains branched at the C(γ) atom (i.e. β^3-HVal) are better helix inducers than those lacking branching at this posititon. The CD spectrum of the β^3-peptide **79** with six β^3-HVal residues is typical of the 3_{14}-helix. However, replacement of β^3-HVal with β^3-HLeu residues in β^3-peptide **80** results in a novel CD spectrum that is no longer characteristic of a 3_{14}-helical structure (compare entries 15 and 16) [168], thus suggesting that the nature of side-chains in β^3-peptides may have a profound effect on their conformational preferences.

Despite the usefulness of CD spectroscopy in the conformational analysis of β-peptides, it is worth bearing in mind at this point that the interpretation of CD spectra requires extreme caution. The CD signature assigned to the 3_{14}-helix might not be unique and unambiguous structural assignement cannot be derived simply from CD measurements. The case of the $\beta^{2,3,3}$-peptide **80** is puzzling [147, 172]. For steric reasons, peptide **80** built from geminally dimethyl-substituted $\beta^{2,3,3}$-amino acid residues, cannot fold into the 3_{14}-helix (the C(a) methyl groups would have to occupy axial positions which is impossible). However, it displays a CD spectrum in MeOH very similar to that of the 3^{14}-helical β^3-peptide **1** which only differs in the intensity of the minimum near 215 nm.

80

The design of β-peptides which can fold into a 3_{14}-helical structure in an aqueous environment is a prerequisite for biological applications. Early studies on β^3-peptides of variable length (up to 20 residues) incorporating two or more polar side chains (e.g. β^3-HLys, β^3-HSer, β^3-HAsp, β^3-HGlu), revealed that in contrast to the situation in MeOH, the CD pattern typical of the 3_{14}-helical structure was barely present in water [97, 98, 168, 173–175]. Various strategies to achieve 3_{14}-helix stabilization in aqueous solution have thus been investigated. One of them consists in the use of various proportions of conformationally constrained cyclic β-amino acids such as trans-ACHC (**40**), trans-DCHC (**41**) and trans-APiC (**42**) [128, 165, 175]. These residues were found to be strong promoters of 3_{14} helicity in aqueous solution. The 3_{14}-helix formed by β-hexapeptide **81** (Fig. 2.14) with alternating ACHC/DCHC residues is remarkably stable in water in the range 25–

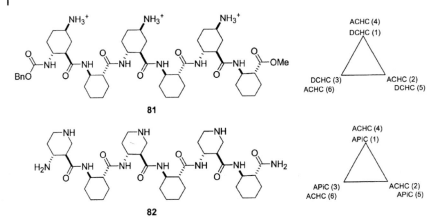

Fig. 2.14 Formulae of β-peptides **81** and **82** forming stable 3_{14}-helical structures in aqueous solution and schematic representation of the position of the amino acid side-chains looking down the 3_{14}-helix axis [128, 165]

80 °C. The CD spectrum recorded in CH_3COOH/CH_3COONa (pH = 3.9, 25 °C) exhibits a strong maximum near 215 nm with a mean ellipticity value approaching 20,000 deg cm^2 dmol^{-1} [165]. The related β-hexapeptide **82** with alternating ACHC/APiC residues also folds into the 3_{14}-helix in water (mean residue ellipticity in 10 mM aqueous Tris, pH = 7 of ca. 10,000 deg cm^2 dmol^{-1}) [128].

It was found subsequently that polar acyclic $β^3$-amino acid residues which offer additional side-chain diversity are tolerated to a certain extent without sacrificing helix stability [175]. Mixed β-nona- and decapeptides containing two-third *trans*-ACHC residues and one-third $β^3$-HLys residues, form extremely stable helical structures in aqueous solution (10 mM aqueous Tris, 150 mM NaCl, pH = 7.2) with mean ellipticity values near 215 nm ranging from ca. 19,000 to as high as 25,000 deg cm^2 dmol^{-1} (e.g. **83**, Fig. 2.15). Reducing the proportion of *trans*-ACHC residues to one-third significantly decreases the ellipticity value at 215 nm (to ca. 7800 deg cm^2 dmol^{-1} in **84**) suggesting that the 3_{14}-helix becomes only partially populated. In the absence of any ACHC residue, the CD spectrum no longer shows the pattern characteristic of the 3_{14}-helix (e.g. **85**, Fig. 2.15). Residual helicity is retained upon insertion of a single conformationally constrained cyclic $β^{2,3}$-amino acid (ACHC, 1,2-dithiane ring) at the central position of $β^3$-hepta- and nonapeptides (e.g. **73**) [108, 168].

Electrostatic interactions between oppositely charged side-chains of α-amino acid residues at positions i and $i+4$ in an α-helix are well known to provide additional helix stability in water. By analogy the creation of ion pairs or salt bridges between complementary charged side-chains of acyclic β-amino acid residues positioned in a $i,i+3$ relationship (on top of each other in the 3_{14}-helix) can stabilize the 3_{14}-helix considerably. This was demonstrated independently by the groups of DeGrado (CD) and Seebach (CD and NMR) who reported that $β^3$-peptides **86** and **87** with salt bridges ($β^3$-HGlu/$β^3$-HOrn or $β^3$-HGlu/$β^3$-HLys) distributed on two faces of the putative 3_{14}-helical structure respectively, fold into a stable helix in aqueous solution

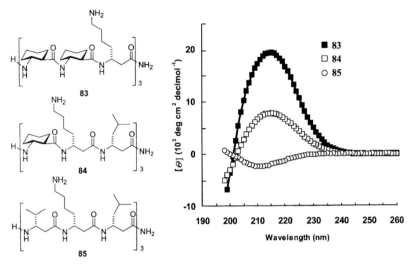

Fig. 2.15 Formulae and CD spectra in aqueous solution of β-peptides with decreasing proportions of ACHC residues from two (**83**) to zero (**85**) [175]

close to neutral pH (where the β-Hlys and β-Orn side chains are expected to be protonated and the β-HGlu side-chains deprotonated) (Fig. 2.16) [175 a, 175 b]. As evidenced by CD measurements, the stability of the helix is strongly affected at extreme pH, thus supporting a crucial role for ion pairing. A similarly strong destabilizing effect is observed upon addition of high concentrations of NaCl (complete loss of ellipticity near 215 nm at 2 M NaCl for **86**) [175 a]. Although NMR studies of these peptides proved to be difficult due to severe spectral overlap, a few interesidue NOEs characteristic of the 3_{14}-helix were unambiguously assigned from the ROESY spectrum of peptide **87** [175 b]. In contrast to peptide **87**, peptide **88** which is obtained by substitution of ACHC residues for the valine residues in **87**, adopted a stable helical conformation independently of ion pairing (no pH effect on helix stability), thus demonstrating that only a few conformationally restrained cyclic *trans*-ACHC residues are sufficient to induce a substantial 3_{14}-helix population in water in an environment-independent manner [176].

As previously discussed, the free NH_3^+ and COO^- termini of β-peptides exert a strong 3_{14}-helix-stabilizing effect mediated by favorable electrostatic interaction with the helix dipoles. In addition, the orientation of the salt-bridge with respect to the helix macrodipole can also contribute significantly to the stability of the 3_{14}-helix. For favorable electrostatic interactions with the helix peptide dipoles to occur, positively-charged side-chains should be placed near the N-terminus and negatively-charged side-chains near the C-terminus (the mean-residue ellipticity of heptapeptide **89** with optimal salt-bridge orientation is twice that of corresponding peptide **87** (Tab. 2.4) [177].

Thus, the optimal design of $β^3$-peptides (e.g. undecamer **90**) which efficiently promote 3_{14}-helicity in aqueous environment can been achieved through a combi-

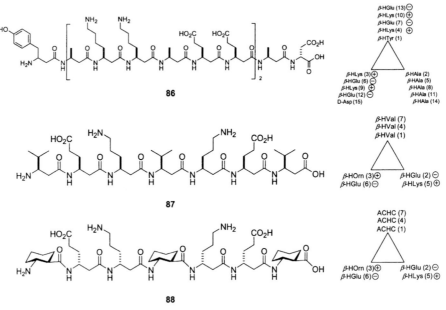

Fig. 2.16 Effect of electrostatic interactions on 3₁₄-helix formation in an aqueous environment [175 a, 175 b, 176]. β^3-Peptides **86** and **87** adopt a stable helical conformation mediated by salt bridges near neutral pH. While the propensity of these peptides to adopt a helical conformation is strongly de- pendent on the pH, helix formation by peptide **88** which incorporates three pre-organized ACHC residues (instead of β-HVal residues in **87**) is hardly affected by pH variations, thus suggesting that it is independent of ion pairing

nation of several features including salt bridge formation between $i/i+3$ side-chains on one helical face, maximization of electrostatic interactions with the helix macrodipole as well as side-chain branching at the C(γ) atom of several amino acid constituents (peptide **91** lacking a branched residue at position 6 displays a much reduced ellipticity at 214 nm compared to **90**) [177].

Since peptides **86–88** and **90** present an amphilic character, the formation of aggregates that could account for the observed increased helicity in water is questionable. Indeed, it was shown by Gellman and coworkers using analytical ultra-centrifugation and NMR spectroscoscopy that an amphiphilic β-decapeptide containing three *trans*-ACHC/*trans*-ACHC/β^3-HLys triads (as in **83**) could self-associate in water at or above 1 mM to form aggregates in the tetramer–hexamer size range [178]. Nevertheless, concentration-dependent CD measurements and sedimentation equilibrium data suggested that under the conditions used (0.2–0.3 mM), peptides **86–88** and **90** were essentially monomeric.

In a first step towards the design of β-peptides tyligomers (oligomers that fold into predictable *tertiary* structures [8]), carefully controlled interhelical hydrophobic interactions have been utilized to stabilize a β^3-peptide two-helix bundle (**92**) [179] (Fig. 2.17).

Tab. 2.4 Effect of salt bridge orientation and side-chain branching at C(γ) on 3_{14}-helicity in water as investigated by CD [a]) [177]

Entry	β-Peptide		$[\theta]_{ca.\ 214\ nm}$
1		89	−8930
2		87	−4370
3		90	−14,250
4		91	−7577

a) CD spectra were recorded in PBC buffer (pH 7, 25 °C) at 0.2 mM peptide concentration. Mean-residue ellipticity $[\theta]$ near 214 nm in deg cm^2 dmol^{-1}.

The bundle was constructed by covalent dimerization of an amphiphilic β^3-peptide through disulfide bond formation. The interaction interface of the bundle consists of hydrophobic residues ((S)-amino valeric acid, β^3-HLeu and β^3-HAla) distributed in a complementary manner on one face of the helix. The 3_{14}-helical fold is stabilized by multiple electrostatic interactions involving charged residues at i and $i+3$ arranged on the two other faces of the helix. The mean-residue ellipticity at 210 nm with a value close to −40,000 deg cm^2 dmol^{-1} indicated that **92** was fully helicoidal. In contrast, the monohelical control (either the reduced peptide or the monomeric sequence with no cysteine residue) exhibited a three-fold lower CD signal. As shown by analytical ultracentrifugation data, peptide **92** was monomeric between 0.28 and 0.8 mM. This result thus indicates that the observed enhanced helicity for **92** can be attributed to intramolecular helix–helix interactions rather than to formation of higher order aggregates [179].

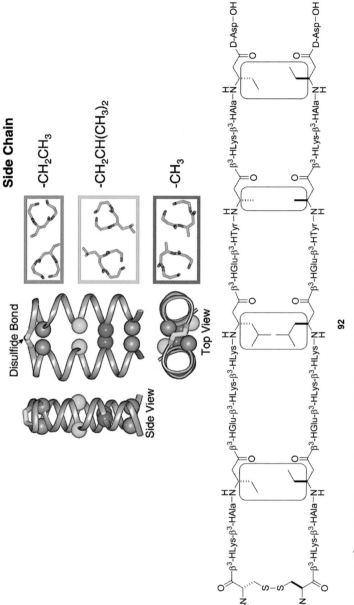

Fig. 2.17 A β^3-peptide **(92)** two-helix bundle [179]. The parallel bundle was designed by dimerizing a 3_{14}-helical peptide via a disulfide bond. The interaction interface of the bundle consist of four hydrophobic residues ((S)-amino valeric acid, β^3-HLeu and β^3-HAla) distributed in a complementary manner on one face of the 3_{14}-helix. Charged side-chains are included to stabilize the helical fold in aqueous solution. Reprinted with permission from [179]. © American Chemical Society (2002)

2.2.3.2 The 12/10- (10/12-) Helix

The conformational preferences of "mixed" β-peptides containing both β^3- and β^2-amino acid residues in their sequence differ markedly from that of the corresponding homopolymers consisting exclusively of β^3- or β^2-amino acid residues. Several types of "mixed" β-peptides have been investigated including *block* peptides constructed with triads of β^3-amino acid residues and triads of β^2-amino acid residues (e.g. **93**) [104,161], as well as *alternating* peptides of β^2/β^3 type (e.g., **72**, **94** and **95**) [104, 161], β^3/β^2 type (e.g. **96–98**)[162], and $\beta^3/\beta^2/\beta^3$ type (e.g. **99**) [104, 161].

Surprisingly, peptides **72** and **93–99** do not display CD spectra in MeOH characteristic of the expected $(M)\text{-}3_{14}$ helix but present a new pattern with an intense single peak near 205 nm (with a mean-residue ellipticity as high as 62,083 deg cm^2 dmol^{-1} for **97**) and no zero crossing (Tab. 2.5).

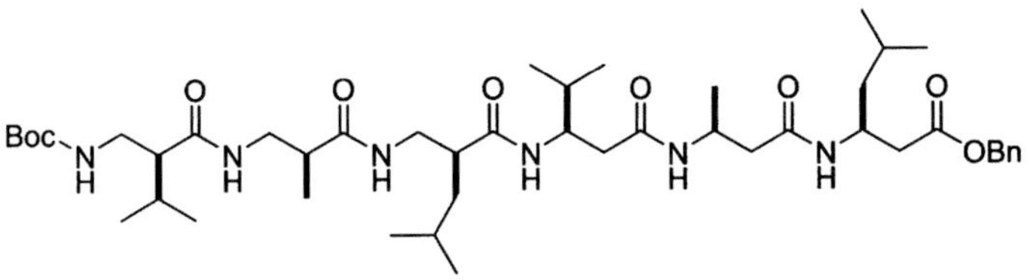

93

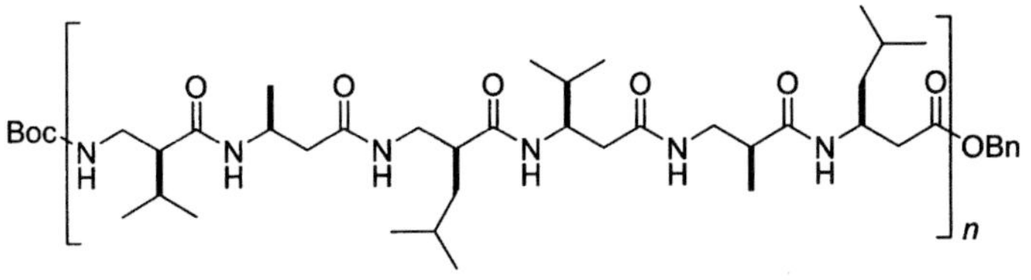

94 *n=1*
95 *n=2*

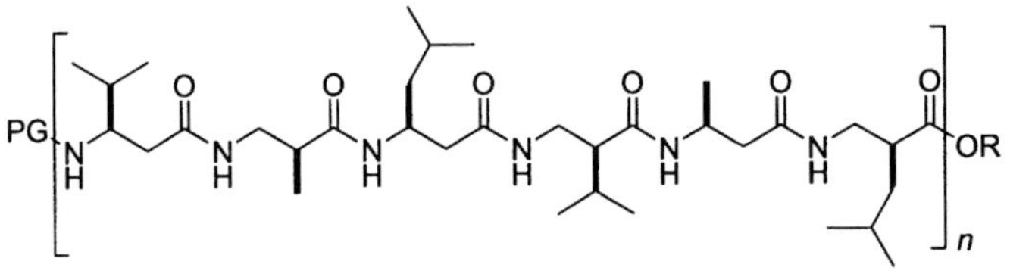

96 *n=1*, PG=Boc, R=Bn
97 *n=2*, PG=Boc, R=Bn
98 *n=2*, PG=H, R=H

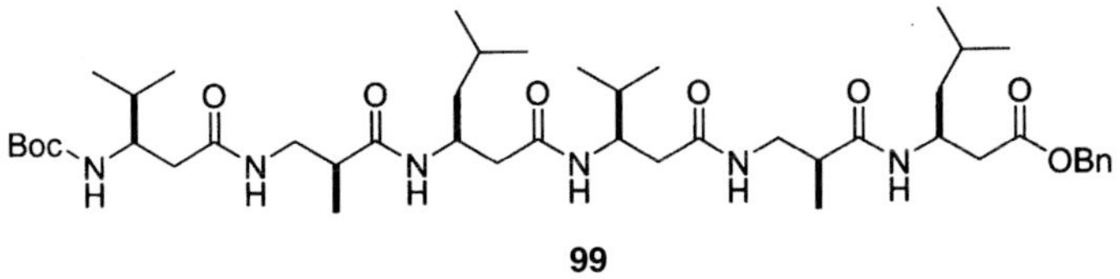

99

Tab. 2.5 Circular dichroism of "mixed" β^2-, β^3-peptides in MeOH at 0.2 mM and mean-residue ellipticity (in deg cm^2 dmol^{-1}) of the maximum at λ_{max}=ca. 203 nm) [104, 161, 162]

	Peptides						
	72	**93**	**94**	**95**	**96**	**97**	**98**
$[\theta]_{\lambda max}$	26,666	20,000	51,666	50,000	58,333	62,083	34,166

This pattern is extremely sensitive to deprotection of the terminal functional groups. Deprotection of both *N*- and *C*-termini results in either a decrease in the maximum ellipticity value at 205 nm (the ellipticity value at 205 nm for **72** is almost two times lower than that of **94**) or a restoration of the CD pattern characteristic of 3_{14}-helicity (upon deprotection of **93** and **95**). More definitive structural information on mixed β-peptides came from detailed NMR analysis of β^2/β^3-peptides **72** (in pyridine and in CD$_3$OH), **94**, and β^3/β^2-peptides **96–98** (in CD$_3$OH). The dispersion of the chemical shifts as well as the large ^3J(H-C(a), H-C(β) values (between 11–12 Hz in the case of **72** in CD$_3$OH) were indicators that at least one stable secondary structure was populated. ROESY experiments revealed a NOE pattern substantially different from that of the 3_{14}-helix with no $i/i+3$ NOE crosspeaks (except **72** for which two $i/i+3$ NOEs absent in pyridine were observed in CD$_3$OH solution) and new $i/i+2$ connectivities not compatible with the 3_{14}-helix. The NOEs and J values were used as distance and dihedral angle restraints respectively, in simulated annealing protocols. All the peptides were found to fold into a novel irregular helical structure with alternating 12- and 10-membered H-bonded rings. The sense of helicity is opposite to that of the corresponding 3_{14}-helix ((P) instead of (M)), and the resulting macrodipole is strongly reduced since C=O and NH amide bonds point alternatively up and down the helix axis. The 12/10-structure of a low energy conformer of β^2/β^3-peptide **72** as determined from NMR measurements in pyridine followed by unrestrained molecular dynamics is shown in Fig. 2.18.

The 12/10-structure is characterized by a central 10-membered H-bonded ring and two 12-membered H-bonded rings at the *N* and *C*-termini. The two turns together with backbone dihedral angles are represented in Fig. 2.19. Comparison of the 10/12-helix turns with the corresponding 14-membered ring of the 3_{14}-helix reveals a common (+)-synclinal arrangement around the central C(a)-C(β) bond for each amino acid constituent. However, whereas both ϕ and ψ angles are negative for β-amino acids in the regular (M)-3_{14}-helix, β^2-amino acid residues have a positive ϕ value and β^3-amino acid residues a positive ψ value in the 12/10-helix.

A similar sequence of 12- and 10-membered turns is present in the structure of Boc-protected β^3/β^2-peptides **96** and **97**, the C=O of the Boc group being engaged in the first 12-membered ring with NH of residue 3. The pattern of 10- and 12-membered turns is reversed for the fully protected β^2/β^3-peptide **94** as well as the unprotected β^3/β^2-dodecapeptide **98** which thus folds into a 10/12-helix, with the NH of residues 1 and 2, respectively being involved in the formation of an *N*-terminal 10-membered turn.

A   **B**

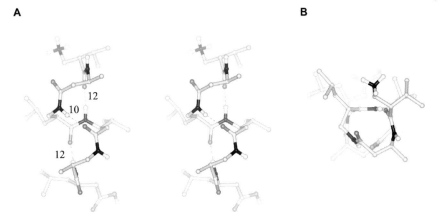

Fig. 2.18 The 12/10-helical structure of mixed β-peptides as determined by NMR measurements in pyridine-d_5. (**A**) Stereo-view along the helix axis of the right-handed 12/10-helix of $β^2/β^3$-peptide **72** obtained by unrestrained molecular dynamics simulation *in vacuo* at 50 K of a low energy conformer resulting from NOE-restrained modeling. (Adapted from [104, 161]). (**B**) Top view

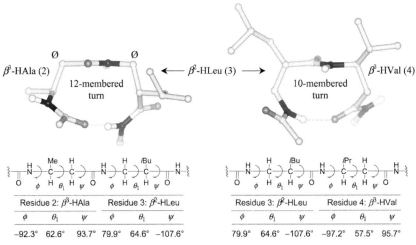

Residue 2: $β^3$-HAla			Residue 3: $β^2$-HLeu		
$φ$	$θ_1$	$ψ$	$φ$	$θ_1$	$ψ$
−92.3°	62.6°	93.7°	79.9°	64.6°	−107.6°

Residue 3: $β^2$-HLeu			Residue 4: $β^3$-HVal		
$φ$	$θ_1$	$ψ$	$φ$	$θ_1$	$ψ$
79.9°	64.6°	−107.6°	−97.2°	57.5°	95.7°

Fig. 2.19 Comparison of the 12- and 10-membered turns found in the 12/10-helix of **72** together with corresponding backbone dihedral angles. The angles were extracted from the low energy conformer of **72** depicted in Fig. 2.18 and based on NMR data

Thus, the tendency is that in the absence of any adjacent substituent on either side of an amide bond, the 12-membered turn is favored, the 10-membered being formed when the amide bond is flanked by substituted carbons. The reduced 12/10-helix population or rearrangement to the 3_{14}-helix observed upon N-terminal deprotection of mixed β-peptides can be explained in terms of unfavorable charge–pole interactions in the right-handed 12/10-arrangment, the positively charged amino terminus being rather a promoter of the left-handed 3_{14}-helical structure (see Section 2.2.3.1).

Recently, a $β^3$-dodecapeptide was found to display a CD spectrum in water which was very similar to that assigned to the 12/10-helix, with a single maximum near 200 nm. Careful NMR analysis however, revealed a predominantely extended conformation without regular secondary structure elements [174]. This result stresses that the CD signature assigned to the 12/10-structure might not be unique and again (see Section 2.2.3.1) that CD spectra must be interpreted with caution.

The formation of the 12/10- (10/12-) helix has been investigated using *ab-initio* quantum mechanics calculations and found to be intrinsically favored over the 3_{14}-helix in the case of unsubstituted β-peptides (oligo-β-HGly peptides) [169, 180, 181]. Reversible folding to the experimentally determined 12/10-helix has also been demonstrated using molecular dynamics calculations (GROMOS96 force field) [105, 166, 182]. Although alternate conformations such as the left-handed 3_{14}-helix (1.3%) were also populated, the right-handed 12/10-helix was the predominant conformation in the simulation of peptide **72** at 340 K.

2.2.3.3 The 2.5$_{12}$-Helix

With a $θ_1$ value close to $±60°$, *trans*-ACHC residues are ideally pre-organized for 3_{14}-helix formation (see Section 2.2.3.1). Varying the ring size will have an immediate effect on the $θ_1$ value and thus on the conformation around the $C(a)$-$C(β)$ bond. This property has been further exploited by Gellman and coworkers to exert rational control over the secondary structure. They predicted that homo-oligomers consisiting of the smalller ring size *trans*-aminocyclopentane carboxylic acid (ACPC) for which the $θ_1$ value is much larger, should form a novel helical structure with ca. 2.5 residues per turn, stabilized by a number of $C=O_i\cdots H-N_{i+3}$ H-bonds closing 12-membered rings [183]. The experimental proof came from detailed conformational analysis in both solution (NMR and CD) and the solid state of hexamer **100** and octamer **101** (Fig. 2.20) with (R,R)-*trans*-ACPC residues [119, 160] which revealed a stable (M)-2.5$_{12}$-helical structure with a pitch of ca. 5.5 Å.

100 *n*=1
101 *n*=2

A

B

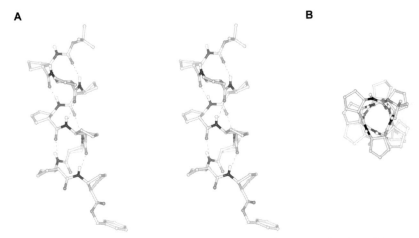

Fig. 2.20 The β-peptide 2.5_{12}-helical structure [119, 160]. **(A)** Stereo-view along the helix axis of the left-handed 2.5_{12}-helix formed by $\beta^{2,3}$-octapeptide **101** consisting of (R,R)-trans-2- amino-cyclopentanecarboxylic acid (trans-ACPC) residues in the solid state as determined by X-ray diffraction. (Adapted from [160]). **(B)** Top view

Examination of the backbone torsion angles in the crystal structure of **101** (Tab. 2.6) shows that for residues 1–7, the θ_1 value is between –87.8 and –113.3° with a mean value of –96° (the somewhat larger θ_1 value for residue 8 is mainly due to fraying of the helix at the C-terminus).

Interestingly, this 2.5_{12}-helical structure and the corresponding 3_{14}-helix of β-peptides with (R,R)-trans-ACHC residues, show opposite polarity and helicity; the 2.5_{12}-helix being polarized from N- to C-terminus like the α-helix. In the crystal, helices are packed through extensive contacts between cyclopentyl rings, with their axes parallel to each other (in a parallel relationship for **100** and in both a parallel and antiparallel relationship for **101**). These lateral interactions between helices are especially relevant in terms of helical bundle formation and the creation of the tertiary structure of β-peptides [119]. As evidenced by NMR studies, the helix is highly populated in organic solvents (pyridine-d_5, CD_3OH) with no evidence of aggregation between 0.25 and 25 mM. Representative interresidue NOEs used to assign the 2.5-helical conformation consist essentially of H-C$(\beta)_i$/NH$_{i+2}$

Tab. 2.6 Backbone torsion angles in the crystal structure of **101** [160]

Angles	Residues							
	1	2	3	4	5	6	7	8
ϕ (°)	83.4	88.9	103.8	108.5	93.1	91.3	115.9	87.2
θ_1 (°)	–102.0	–87.8	–97.8	–92.5	–92.2	–89.1	–113.3	–155.1
ψ (°)	121.9	110.6	98.7	97.8	112.2	109.9	93.1	–

and H-C(β)$_i$/H-C(a)$_{i+2}$ together with weak H-C(β)$_i$/NH$_{i+3}$ [159,160]. In MeOH, hexamer **100** exhibits a typical CD signature with a maximum at 204 nm, zero-crossing at about 214 nm and a minimum at 221 nm, and this differs substantially from the CD spectra of 3_{14}-helical β-peptides (see Section 2.2.3.1) [119, 160]. This pattern agrees well with theoretical CD spectra calculated for helical poly(*trans*-2-aminocyclopentanecarboxylic acid) [184].

With the aim of studying the formation of the 2.5-helix in water as well as to introduce side chain diversity, Gellman and his group synthesized β-peptides from a variety of β-amino acid building blocks (**34–39**) constrained with five-membered rings (Fig. 2.21).

Mixed β-peptides consisting of ACPC residues and aminopyrrolidine carboxylic acid residues (*trans*-APC or *trans*-AP residues; e.g. **102**, **103**) or ACPC and hydrophilic 3-substituted ACPC residues (e.g. **104**) with as few as six residues, adopt a robust 2.5-helical conformation in aqueous solution [121, 123–125]. The CD spectra of these peptides recorded in water exhibit a strong maximum near 201 nm with mean ellipticity values ranging from ca 10,000 to 15,000 deg cm^2 dmol^{-1}, typical of the 2.5-helical structure. The intensity of the extrema at 201 and 221 nm increases in a length-dependent manner with an isodichroic point at 213 nm suggesting that helix stability increases as the peptide becomes longer. NMR analysis of **102** in 9:1 H$_2$O:D$_2$O (100 mM acetate, pH 3.8) was more challenging than in CD$_3$OH due to signal overlaps, however, the characteristic H-C(β)$_i$/NH$_{i+2}$ and H-C(β)$_i$/H-C(a)$_{i+2}$ NOE pattern appeared to be maintained. Given the geometrical similarities (pitch, dipole orientation, helicity) with the a-helix, the 2.5-helix might serve as a scaffold to create rationally designed mimetics of helical a-peptides capable of interfering with protein–protein interactions. Several strategies have thus been developed to achieve additional and specific functionalization of 2.5-helical β-peptides without sacrificing helix propensity in water. In particular β-peptides containing sulfonylated APC residues (e.g. **105**), and 3-substituted ACPC residues (e.g. **104**, **106**) have been prepared and evaluated for helix formation in an aqueous environment [123, 125]. Bidimentional NMR analysis of **105** and **106** in 9:1 H$_2$O:D$_2$O (4 °C) revealed that both peptides display a interresidue NOE pattern (H–C(β)$_i$/NH$_{i+2}$, H-C(β)$_i$/H-C(a)$_{i+2}$ and H-C(β)$_i$/NH$_{i+3}$ connectivities) characteristic of the 2.5_{12}-helical conformation. No NOEs inconsistent with the 2.5-helical structure were observed. Similarly, CD spectra of peptides **104** and **105** in water were consistent with a 2.5-helical conformation and compared well with that of peptide **102**. The maximum near 200 nm for **105** is more intense in MeOH and slightly shifted to the red compared to H$_2$O (Fig. 2.22), thus suggesting enhanced helicity in MeOH relative to H$_2$O.

The finding that substantial 2.5-helicity may be retained in water upon the introduction of a limited number of acyclic β^3- or β^2-amino acid residues at chosen positions in the sequence, further expand the side-chain array available for functionalization of the 2.5-helical scaffold. In water, β-heptapeptides **107** and **108** which contain two β^3-amino acid residues [184a] and two β^2-amino acids [184b], respectively, still display a CD spectrum and NOE connectivities characteristic of the 2.5-helix. However, the addition of a third acyclic amino acid is detrimental to the formation of the 2.5-helix in water.

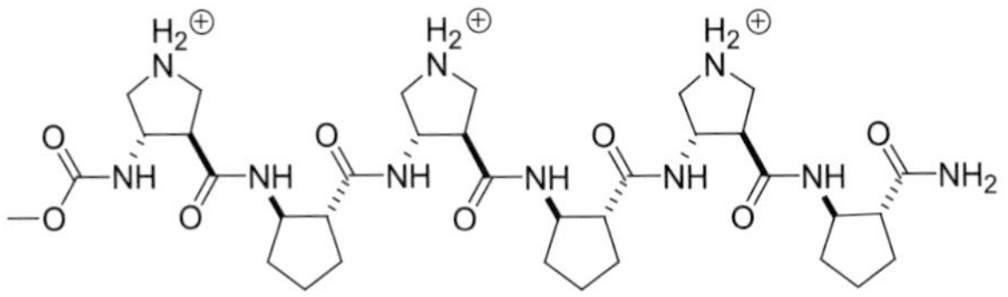

102

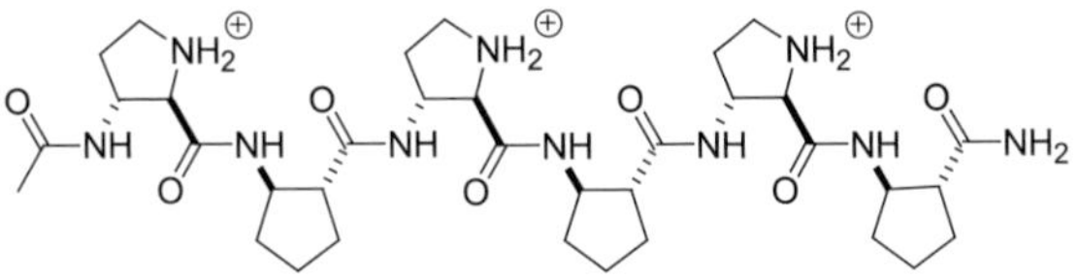

103

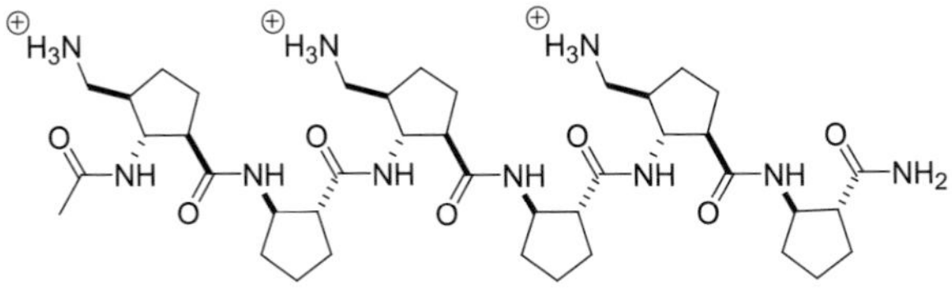

104

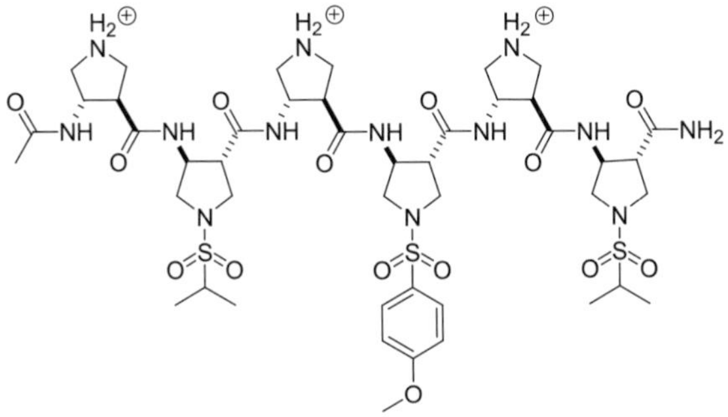

105

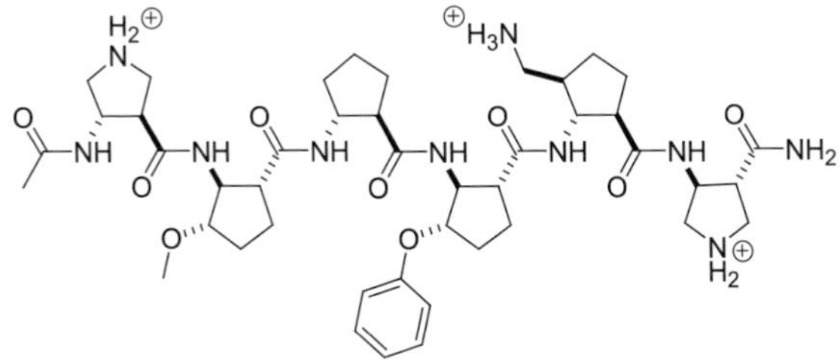

106

Fig. 2.21 Water soluble β-peptides incorporating β-amino acid residues constrained with five-membered rings that fold into the 2.5_{12}-helical secondary structure [121, 123–125]

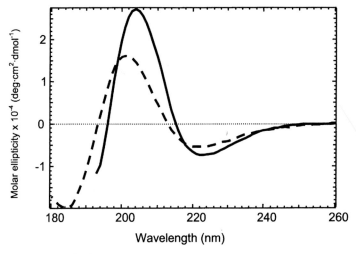

Fig. 2.22 Circular dichroism spectra for β-peptide **105** (0.1 mM, 25 °C) in MeOH (solid line) and water (dashed line). The vertical axis is the mean-residue ellipticity [123]. Reprinted with permission from [123]. © American Chemical Society (2001)

107

108

Of course, the number of acyclic residues tolerated in 2.5-helical peptides will vary as a function of the peptide length. For example, CD measurements performed on an amphiphilic 17-mer peptide with ACPC and APC residues and containing six β³-Hleu residues, shows the expected signature in water with a mean-residue ellipticity at ca. 202 nm close to 11,000 deg cm^2 dmol^{-1} (albeit weaker than in MeOH for which $[\theta] > 30,000$ deg cm^2 dmol^{-1}) [184a].

2.2.3.4 The 2$_8$-Helix

For steric reasons, β-peptides consisting of geminally disubstituted amino acids or of β^{2,3}-amino acids of *unlike* configuration do not fit into any of the three helical folds described previously. Surprisingly, while all-*unlike*-β^{2,3}-peptides with alkyl substituent at both α- and β-positions form extended strands that arrange into pleated sheets (*vide infra*, Section 2.2.4), analogous peptides with α-hydroxy substituents have been found to adopt a helical conformation reminiscent of the 2$_1$-helix of PHB (see Section 2.2.1, Fig. 2.3) [111]. The conformation of β-hexapeptide **109** was investigated by NMR in CD$_3$OH solution.

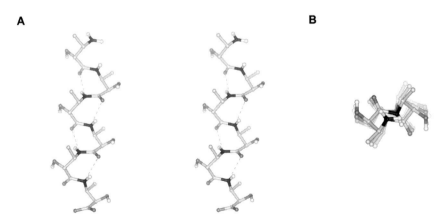

109

Proton resonances for all residues were assigned using a combination of COSY, TOCSY, HSQC, and HMBC experiments. The large values of $^3J(NH/H\text{-}C(\beta))$ and the small values of $^3J(H\text{-}C(\alpha)/H\text{-}C(\beta))$ were indicative of antiperiplanar and syn-clinal arrangements respectively, around those bonds. In addition, medium-range NOE connectivities H-C(β)$_i$/NH$_{i+1}$, H-C(α)$_i$/NH$_{i+1}$, NH$_i$/NH$_{i+1}$ were consistent

A   **B**

Fig. 2.23 Model of the 2$_8$-helix formed by all-*unlike*-β^{2,3}-peptide **109** generated with ideal torsion angle values $\phi = -135°$, $\theta_1 = 58°$, $\psi = 27.5°$ derived from NMR meaurements in CD$_3$OH. (Adapted from [111]). (**A**) Stereo-view along the helix axis. (**B**) Top view

with the presence of a compact secondary structure. The structure determined by molecular modeling under NMR restraints is a right-handed helix (a model constructed with "ideal" backbone torsion angles $\phi=-135°$, $\theta_1=58°$, $\psi=27.5°$ is shown in Fig. 2.23) with approximately two residues per turn and a pitch of 4.7–4.9 Å, held by $C=O_i \cdots H-N_{i+2}$ H-bonds closing eight-membered pseudocycles.

In addition, the nearly *synperiplanar* arrangement of the a-hydroxyl and carbonyl groups (torsion angle of ca. −32°) suggests that $C=O_i \cdots H-O_i$ H-bonds could also play a significant role in inducing and stabilizing the 2_8-helical fold. Furthermore, examination of crystal structures of taxol derivatives [185, 186] and other molecules incorporating an a-hydroxylated β-amino acid [187–189] revealed a similar arrangement around the CO-C(a) bond of the a-hydroxylated $β^{2,3}$-amino acid residue. This observation is consistent with pre-organization at the monomeric level towards helix formation.

Interestingly, the 2_8-helical fold identified by NMR analysis of β-peptide **109** compares well with the model of a $β^{2,2}$-peptide consisting of 1-aminomethylcyclopropanecarboxylic acid residues (Fig. 2.24). This model was generated using ideal torsion angle values ($\phi=+120°$, $\theta_1=-72°$, $\psi=0°$, and $\omega=180°$) derived from crystal structures of dimer **110**, trimer **111** and tetramer **112** [163] (Fig. 2.25).

These structures are all characterized by the presence of eight-membered H-bonded rings between $C=O_i$ and $H-N_{i+2}$. Evidence for intramolecular H-bonding in solution came from FT-IR analysis of the NH stretch region in the FT-IR spectra. As expected the ratio of bonded (NH stretch at 3285–3344 cm^{-1}) to non-bonded NH (NH stretch at 3446–3456 cm^{-1}) groups was found to increase gradually with chain length. 1-Aminomethylcyclopropanecarboxylic acid residues significantly restrict peptide backbone conformations because of an enlarged exocyclic bond angle C(β)-C(a)-CO and because of the so-called bisecting effect [190] which locks the ψ torsion angle at a value close to 0° (Fig. 2.24). Interestingly, when going from the cyclopropane- to the cyclohexane-based monomer **57**, there is no longer a preference for the bisected conformation and peptides consisting of 1-aminomethylcyclohexanecarboxylic acid residues display a different conformational preference [145]. The crystal structure of β-tripeptide **113** (Fig. 2.26) is char-

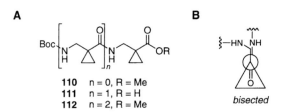

A

110   n = 0, R = Me
111   n = 1, R = H
112   n = 2, R = Me

B

bisected

Fig. 2.24 (A) Sequence of β-peptides containing 1-(aminomethyl)cyclopropanecarbonyl moieties studied by X-ray diffraction. These peptides were synthesized in solution using **58** and the related methyl ester as building blocks [163]. (B) Bisected conformation around the exocyclic C-CO bond observed in X-ray crystal structures of peptides **110–112**

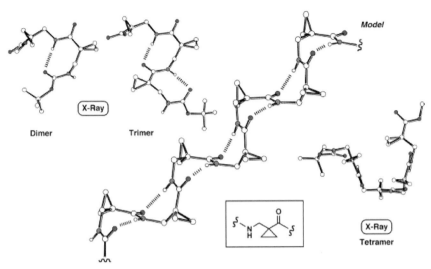

Model

X-Ray

Dimer   Trimer

X-Ray

Tetramer

Fig. 2.25 Conformational preferences and eight-membered turn motif of $\beta^{2,2}$-peptides with 1-aminomethylcyclopropanecarboxylic acid residues. X-ray crystal structures of dimer **110**, trimer **111** and tetramer **112** together with a model constructed using slightly modified torsion angles ($\phi=+120°$, $\theta_1=-72°$, $\psi=0°$, and $\omega=180°$) derived from X-ray crystal structure of **111** (generated with MacMoMo). View adapted from [163] and kindly provided by Professor Dr Seebach (ETH, Zürich)

acterized by a 10-membered H-bonded turn segment involving the NH of residue 2 and the C=O group of residue 3 which show structural similarity to the central 10-membered turn of the 12/10/12-helix (see Section 2.2.3.2. and Fig. 2.19). The backbone torsion angles for the 10-membered turn in **114** are as follows: residue 2 ($\phi=101.8°$, $\theta_1=59.6°$, $\psi=75.9°$); residue 3 ($\phi=93.6°$, $\theta_1=-64.8°$, $\psi=-81.8°$).

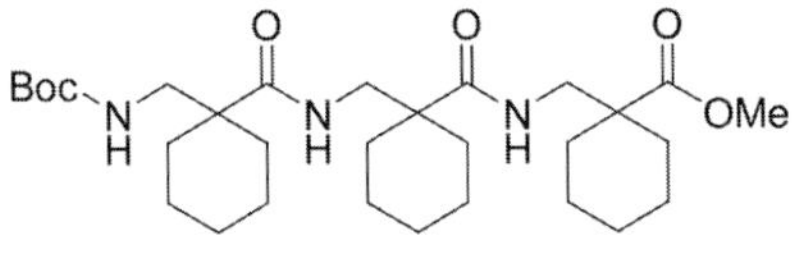

113

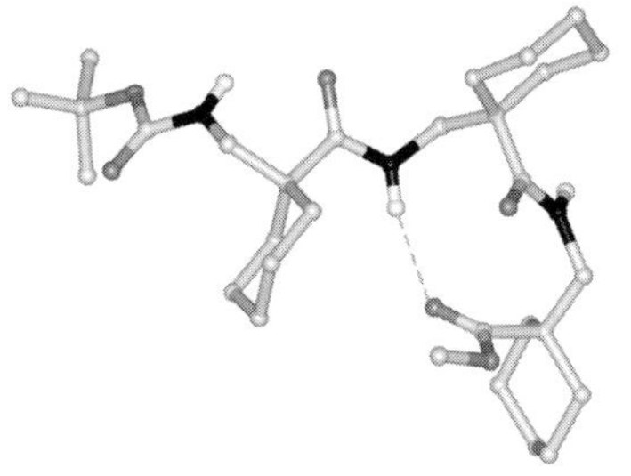

Fig. 2.26 The crystal structure of $\beta^{2,2}$-tripeptide **113** with 1-aminomethylcyclohexanecarboxylic acid residues reveals a 10-membered H-bonded turn segment involving NH of residue 2 and C=O of residue 3 [145]. The turn segment shows structural similarity to the central 10-membered turn of the 12/10/12-helix (see Section 2.2.3.2 and Fig. 2.19). The intramolecular H-bond is characterized by a N⋯O distance of 3.10 Å and a (N-H⋯O) angle of 169.4°

2.2.4

Extended β-Peptide Strands, Turns and Formation of Sheet Structures

As mentioned in the introductory part of Section 2.2.3, fully extended β-peptide strands can be generated by populating antiperiplanar conformations around the $C(a)$-$C(\beta)$ bond (θ_1 values close to 180°; Fig. 2.27A).

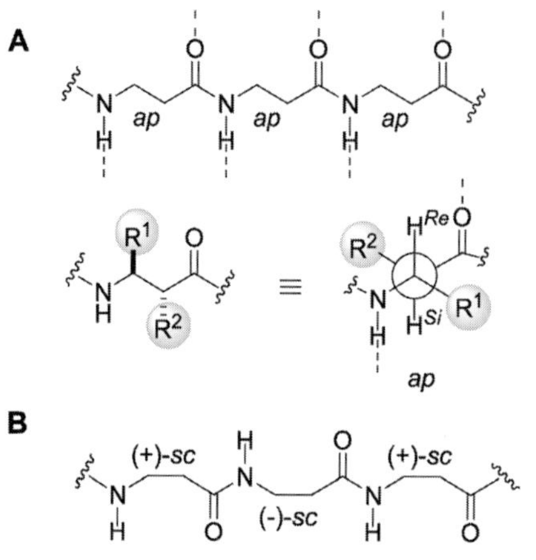

A

B

Fig. 2.27 The two types of extended β-peptide strands with conformation requirements around the $C(a)$-$C(\beta)$ bonds. (**A**) Parallel and antiparallel polar sheets with antiperiplanar conformations around the $C(a)$-$C(\beta)$ bond are promoted by unlike-$\beta^{2,3}$-amino acids with alkyl side-chains. Antiperiplanar side-chains at $C(a)$ and $C(\beta)$ occupy positions approximately perpendicular to the amide planes. (**B**) Extended strands formed by alternating (+)-sc and (−)-sc conformations

Seebach's and Gellman's group found that this conformational bias can be introduced by using acyclic $\beta^{2,3}$-amino acids of *unlike* configuration bearing alkyl substituents [109, 191]. The resulting $\beta^{2,3}$-peptide chains adopt extended conformation with formation of pleated sheets. This is in contrast to all-*like*-$\beta^{2,3}$-peptides which have be shown to form predominantely 3_{14}-helical structures (see Section 2.2.3.1). A second type of extended conformation consisting of alternating (+)- and (−)-*synclinal* arrangements around $C(a)$-$C(\beta)$ bonds (Fig. 2.27 B) has also been described for β-peptide chains built from β-HGly (β-alanine) [109, 154] and *unlike*-$\beta^{2,3}$ amino acids **23** (e.g. **114**) [110] and **44** (e.g. **115**) [130].

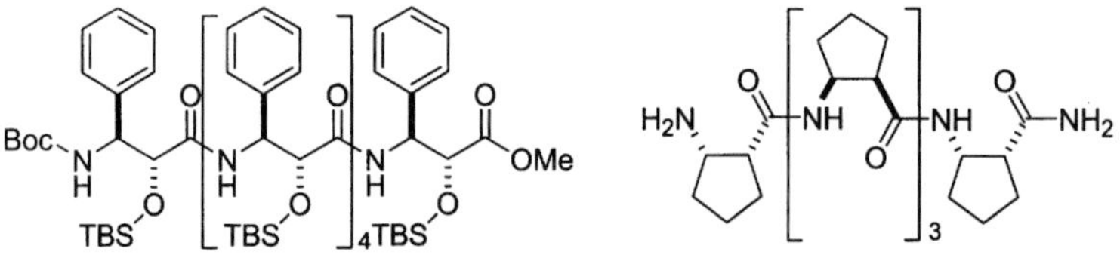

114

115

A β-peptidic parallel sheetlike conformation with characteristic intermolecular 14-membered H-bonded rings was first observed in the crystal structure of β^3-peptide Boc-β^3-HVal-β^3-HAla-β^3-HLeu-OMe **116** (Fig. 2.28 A) [10]. However, the structure also displays a turn which can be considered as the starting point of a 3_{14}-helix. The corresponding all-*unlike*-$\beta^{2,3}$-peptide **117** with methyl substituents at α-positions was prepared to enforce the parallel pleated sheet arrangement. The X-ray crystal structure of **117** (Fig. 2.28 B) reveals the expected fully extended backbone conformation and the antiperiplanar arrangement of alkyl substituents at C(α) and C(β) [191]. The resulting parallel pleated sheet structure which is highly polar with all C=O and NH groups pointing respectively, in opposite directions, differs fundamentally from α-peptide β-sheets where the C=O of neighboring residues point in the opposite direction thus resulting in apolar sheet. The polarity of the stacking might account for the very low solubility of *unlike*-$\beta^{2,3}$-peptides.

To prevent insolubility resulting from uncontrolled aggregation of extended strands, two adjacent parallel or antiparallel β-peptide strands can be connected with an appropriate turn segment to form a hairpin. The β-hairpin motif is a functionally important secondary structural element in proteins which has also been used extensively to form stable and soluble α-peptide β-sheet arrangements in model systems (for reviews, see [1, 4, 5] and references therein). The need for stable turns that can bring the peptide strands into a defined orientation is thus a prerequisite for hairpin formation. For example, type I' or II'' turns formed by D-Pro-Gly and Asn-Gly dipeptide sequences have been found to promote tight α-peptide hairpin folding in aqueous solution. Similarly, various connectors have been

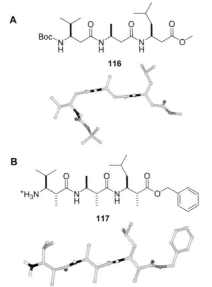

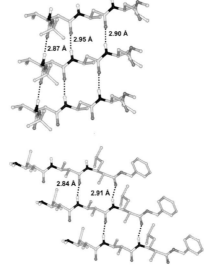

Fig. 2.28 X-ray crystal structures of parallel sheet-forming β³- and all-*unlike*-$\beta^{2,3}$-peptides **116** and **117** [10, 191]. Views along the parallel amide planes and crystal packing diagram show the parallel pleated sheet arrangement (view perpendicular to the amide planes).

(A) β³-Peptide **116**. Only one of the two independent molecules present in the crystal is shown. The NH⋯O angles are 163.3, 168.4 and 168.3°. **(B)** All-*unlike*-$\beta^{2,3}$-peptide **117**. The NH⋯O angles are 159.7 and 165.6° respectively

designed to enforce the formation β-peptide antiparallel pleated sheets (Figs. 2.29 and 2.30).

These include the depsipeptide segment L-Pro-glycolate (e.g. **118**) [109], the α-peptide D-Pro-Gly sequence (e.g. **119**) [192], 12-membered heterochiral dinipecotic (Nip or β²-HPro) turn segments (e.g. **120–121** with (R)-Nip-(S)-Nip sequence) [154, 193], and the 10-membered turn formed by mixed β²/β³-dipeptides (see Section 2.2.3.2; Fig. 2.19; e.g. **122**) [191, 194]. X-ray crystal structures of peptides **118**, **119**

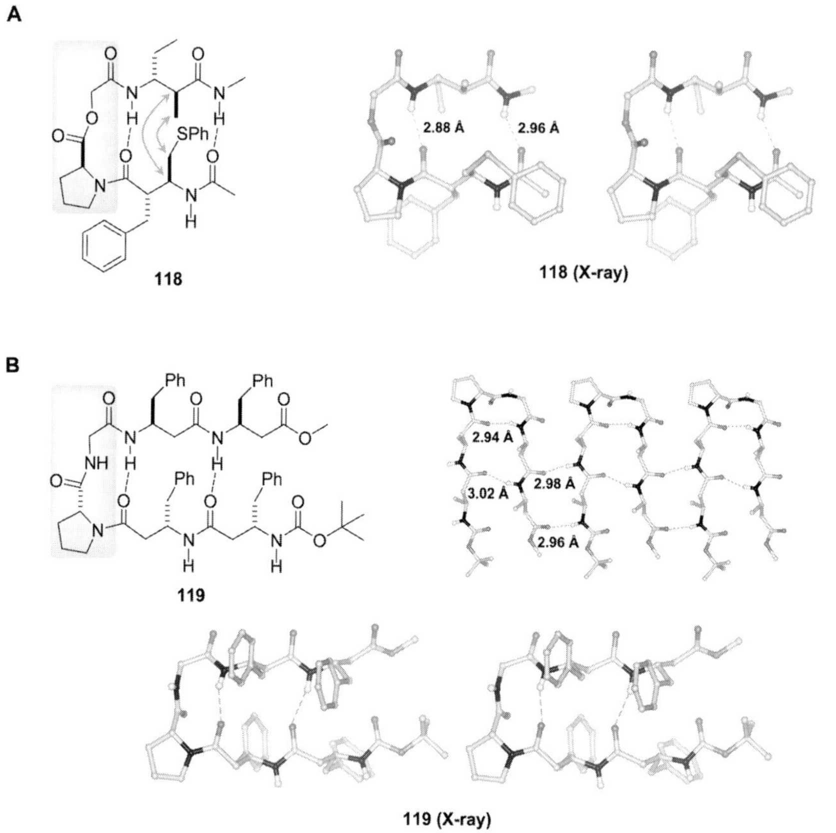

Fig. 2.29 Comparison of antiparallel hairpins **118** and **119** with non–peptide turn segments. (A) Hybride β-peptide **119** with an L-Pro-glycolate turn segment (gray color): summary of backbone–backbone and side-chain–side-chain NOEs (NMR in CD₂Cl₂ and CD₃OD) as well as X-ray crystal structure (stereo-view) [109]. Two independent molecules with similar conformations are present in the crystal; only one is represented. The intramolecular H-bond N···O distances are shown. The angles (N-H···O) are 149.7° (inner H-bond) and 144.7° (outer H-bond). (B) Hybride β-peptide **120** with a D-Pro-Gly type II′ β-turn segment (gray color): X-ray crystal structure [192]. The intramolecular H-bond N···O distances are shown. The angles (N-H···O) are 147° (inner H-bond) and 155° (outer H-bond). The intermolecular NH···O=C H-bonds (with N-H···O angles of 160 and 133°) connect the hairpin into an infinitely extended β-sheet

(Fig. 2.29 A and B) and **121** (Fig. 2.30 A, see p. 80) show the expected antiparallel hairpin conformation, all strand residues displaying an antiperiplanar arrangement around the C(*a*)-C(*β*) bond [109, 192, 193]. Similarly, the nipecotic acid residues in **121** also display θ_1 values close to 180°. The tertiary amide between the two Nip residues has an *E* configuration while the other tertiary amide possesses the *Z* configuration. These hairpin structures are characterized by the unidirectionality of the C=O and NH bonds within each strand segment. In the crystal structure of **119**, hairpins form strong unidirectional C=O⋯NH hydrogen bonds with neighboring molecules that generate infinite polar pleated -sheets with adjacent hairpin molecules parallel to each other [192]. Detailed spectroscopic studies (FT-IR and NMR) of hybride peptide **118** and *β*-peptides **120** and **122** revealed a large hairpin population in non-polar (CD$_2$Cl$_2$) and/or polar solvents (CD$_3$OH, CD$_3$OD). A bundle of lowest energy conformers of hairpin **122** generated under NMR restraints from studies in CD$_3$OH [191, 194], is shown in Fig. 2.30 B. Interestingly, the 10-membered turn segment in **122** shows a certain structural similarity with the type II′ *β*-turn of *a*-peptides (Fig. 2.31).

NMR analysis of peptide **122** provided no evidence for secondary structure in H$_2$O/D$_2$O [195]. Measurements were conducted at low dilutions to preclude aggregation and intermolecular H-bonding. Concentration-dependent NMR studies conducted on peptides **118** and **120** in CD$_2$Cl$_2$ confirmed that little aggregation occurs below 5 and 1 mM respectively. In non-polar solvents and at low concentration, FT-IR spectra peptides **118** and **120** display intense H-bonded NH-stretch bands indicative of extensive intramolecular hydrogen bonding. To determine the influence of the stereochemistry on turn formation and hairpin structure, diasteomers of **120** with *S/R*, *S/S* and *R/R* configuration at the dinipecotic acid turn segment were also synthesized and studied by FT-IR and NMR (amide proton chemical shift). The *S/R* segment was found to promote efficient intramolecular hydrogen bonding albeit to a lower extent compared to the *R/S* segment. The reason could be that while a single rotameric form largely predominates in the *R/S* segment (94:6 rotamer ratio), the minor rotamer is more largely populated in the *S/R* segment (78:22 rotamer ratio). Neither peptide with homochiral dinipecotic acid turn sequences allowed the formation of the 12-membered H-bonded ring.

$\beta^{2,3}$-Amino acids in the strands are typically characterized by large *J*(H-C(*a*),H-C(*β*)) values consistent with an antiperiplanar arrangement around the C(*a*)-C(*β*) bond [109, 191]. In addition all these peptides display a characteristic NOE pattern [109, 191–194] (see summary of observed long-range NOEs in Figs. 2.29 and 2.30) consisiting of (i) long-range interstrand backbone NOE connectivities between H-C(*β*) on the H-bonded acceptor strand and H-C(*a*) on the H-bonded donor strand (H-C(*a*)$_2$/H-C(*β*)$_5$ in the case of **122**; H-C(*β*)$_1$/H-C(*a*)$_4$ in **118** and **120**) and (ii) long-range interstrand side chain–side chain NOEs between the C(*a*)-substituent on the H-bonded acceptor strand and the C(*β*) substituent on the H-bonded donor strand for **120** and **122**. A reverse NOE pattern between C(*β*)-substituent on the H-bonded acceptor strand and the C(*a*) substituent on the H-bonded donor strand is observed for **118**.

A

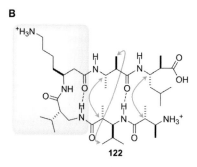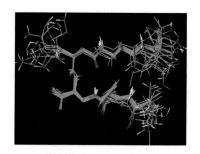

120 R = Me
121 R = OtBu

121 (X-ray)

B

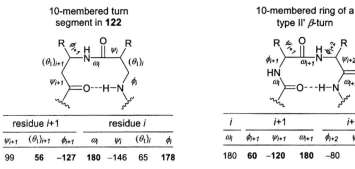

122

Fig. 2.30 Comparison of antiparallel hairpin structures in *β*-peptides **120–122**. (**A**) *β*-Peptides **120, 121** with a 12-membered *R/S* dinipecotic (Nip or *β*²-HPro) turn segment (gray color). Summary of backbone–backbone and side-chain–side-chain NOEs collected in CD₂Cl₂ and X-ray crystal structure of **121** (stereo-view) [154, 193]. The intramolecular H-bond N···O distances are shown. The angles (N-H···O) are 170.8° (inner H-bond) and 172.3° (outer H-bond). (**B**) *β*-Peptide **122** with the 10-membered turn formed by a mixed *β*²/ *β*³-dipeptide sequence: *β*²-HVal-*β*³-HLys [191, 194]. Summary of backbone–backbone and side-chain–side-chain NOEs collected in CD₃OH and superimposition of the 15 lowest energy conformers (taken from [191]) from *ab-initio* simulated–annealing runs with experimental restraints derived from the NMR data (NOE and *J* values). The central portion of the hairpin is well defined but the *N*- and *C*-terminal residues 1 and 6 are more flexible

10-membered turn
segment in **122**

residue *i*+1					residue *i*	
ψ_{i+1}	$(\theta_1)_{i+1}$	ϕ_{i+1}	ω_i	ψ_i	$(\theta_1)_i$	ϕ_i
99	56	**−127**	**180**	**−146**	65	**178**

10-membered ring of a
type II' *β*-turn

i	*i*+1			*i*+2		
ω_i	ϕ_{i+1}	ψ_{i+1}	ω_{i+1}	ϕ_{i+2}	ψ_{i+2}	ω_{i+2}
180	60	−120	180	−80	0	180

Fig. 2.31 Comparison of the turn segment found in hairpin **122** with a naturally-occuring type II' *β*-turn of *α*-polypeptides together with backbone dihedral angles in degrees. In the case of **122**, the angles were extracted from one low energy conformer derived from NMR data and shown in Fig. 2.30. Torsion angles with comparable values are shown in bold [191, 195]

Tab. 2.7 Comparaison of selected backbone torsion angles for strand segments in antiparallel sheet-forming β-peptides **117–119, 121, 122** [109, 154, 191–194]

β-Peptides									
117[a]		**118**[a, b]		**119**[a]		**121**[a]		**122**[c]	
Residue 1	Residue 2	Residue 1	Residue 4	Residue 2	Residue 5	Residue 1	Residue 4	Residue 2	Residue 5
ϕ (°) −107	−116	134	111	−113	−84	94	130	−118	−110
θ_1 (°) 177	178	−179	175	153	171	−165	173	160	173
ψ (°) 127	127	−118	−133	110	115	−122	−135	66	148

a) X-ray crystal structure.
b) Data taken from one of the two independent molecules present in crystals of **119**.
c) From a low energy conformer of β-peptide **122** generated by molecular modeling under NMR-restraints [195].

As a consequence of their different turn geometry: a 10-membered turn closed by H-bonds between NH_i and $C=O_{i+1}$ and a 12-membered turn closed by H-bonds between $C=O_i$ and NH_{i+3}, antiparallel hairpins formed by β-peptides **121** and **122** display opposite sheet polarities (see Fig. 2.30 A and B). Comparison of backbone torsion angles (X-ray and NMR) for selected β-amino acids residues within extended strand segments of peptides **117–122** are shown in Tab. 2.7. The observed values are close to ideal values for β-peptide pleated sheets: $\phi = -120°$ (or 120°), $\theta_1 = 180°$, $\psi = 120°$ (or −120°).

As a result of their capacity to tolerate very different kinds of β-peptidic turns (**120** versus **122**) and different stereochemistries in the turn segment (compare S/R and R/S nipecotic acid dipeptide sequences), hairpin structures formed by β-peptides reveal unusually high plasticity compared to α-peptide β-hairpins.

Furthermore, D- and L-prolyl-(1,1-dimethyl)-1,2-diaminoethyl (Pro-DADME) segments have been used successfully to connect two β-peptide strands via their C-termini and to promote the formation of a parallel hairpin (e.g. **123**) in both solution (CD_3OH) and solid state (Fig. 2.32) [196].

In recent work, Balaram and coworkers have demonstrated that β-amino acids can be tolerated at specific positions in the strands and turn segments of α-peptide hairpins [197, 198]. Induction of turn structure in tetrapeptides by incorporation of a $\beta^{2,3}$-amino acid has also been described [115]. Similarly, studies by Gellman have shown that the dinipecotic acid heterochiral turn segment can be used to nucleate α-peptide hairpin structures [199].

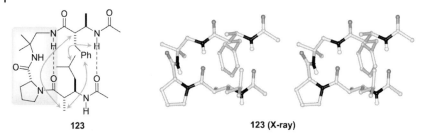

123   123 (X-ray)

Fig. 2.32 Parallel hairpin formation in a β-peptide by incorporation of a D-Pro-DADME turn segment [196]. Summary of the long-range NOE observed by NMR in CD₃OH which is consistent with a hairpin conformation together with the structure of **123** in the solid state. The intramolecular H-bond N···O distances are 2.98 Å (inner H-bond) and 2.85 Å (outer H-bond). The angles (N-H···O) are 160.5° (inner H-bond) and 156.6° (outer H-bond)

2.3
Molecular Organization in γ-Peptide Oligomers

Homologs of β-peptides with one additional methylene group inserted into the backbone of each residue, namely γ-peptides (Fig. 2.33), also form well-defined

Substitution patterns in γ-peptides that lead to defined secondary structures

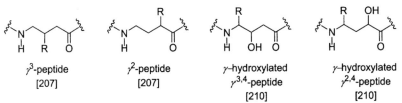

γ⁴-peptide [200,201]   γ²,⁴-peptide [201-205]   γ²,³,⁴-peptide [206-207]

vinylogous-peptides [208,209]

γ-Peptides with no preferred or still undetermined secondary structure

γ³-peptide [207]   γ²-peptide [207]   γ-hydroxylated γ³,⁴-peptide [210]   γ-hydroxylated γ²,⁴-peptide [210]

Fig. 2.33 Classification of γ-peptides according to their substitution patterns (Seebach's nomenclature) and folding propensity

secondary structures (helix, turns, and sheets) in solution and/or in the solid state [200–210].

The first suggestion that a peptide chain consisting of γ-amino acid residues could adopt regularly folded structures was made 40 years ago from optical rotatory dispersion studies in acid solution of the naturally occurring poly(γ-glutamic acid) (PGGA), a high molecular weight polymer produced by bacteria of the genus *Bacillus* [211]. Although a helical fold was suspected (two models characterized by 17- and 19-H-bonded pseudocycles have been proposed, the latter of which being supported by recent atomic-resolution computer simulations [212]), considerable differences in the structure were observed depending on the degree of ionization of the polymer and the detailed structure adopted by PGGA has not been elucidated so far. Only recently, the structures in the solid state of the related poly(α-alkyl γ-glutamates)s were investigated and more definitive models (a 2.5_{14}-helix, a 3.7_{19}-helix and a sheet structure) derived from X-ray diffraction and molecular modeling data [213, 214].

The finding that intramolecular H-bonding occurs in model peptides derived from γ-aminobutyric acids (e.g. **11**) suggested that H-bonds between nearest neighbors in γ-peptides (C7 and C9 folding patterns) could eventually compete with long-range H-bonds, thus preventing the formation of defined secondary structures [63]. However, pioneering work by Seebach [200, 203, 206] and Hanessian [201, 202] has demonstrated that pre-organization of the main chain for long-range H-bonds is favored in γ-peptides as short as four residues with appropriate substitution pattern. The intrinsic conformational preferences of substituted γ-amino acids derives in part from the avoidance of destabilizing *syn*-pentane interactions [215, 216]. In the case of a 2,4-substitution pattern, only two out of nine conformations generated by rotation around $C(α)-C(β)$ and $C(β)-C(γ)$ bonds are free of *syn*-pentane interaction (versus five out of nine for monosubstituted $γ^4$-amino acids) (Fig. 2.34).

This effect is particularly well documented for $γ^{2,4}$- and $γ^{2,3,4}$-amino acid residues [217, 218] which in several natural products (bleomycin A_2 [219], calyculins [220]) have been shown to play a substantial role in the pre-organization of the whole molecule into its bioactive conformation. For example, changes in the substitution pattern of the γ-amino acid linker in bleomycin A_2 result in reduced DNA cleavage efficiency [219]. In the case of γ-peptides, changing the relative configuration *like* or *unlike* of $γ^{2,4}$-amino acids has been used as a strategy to generate different local conformations (Fig. 2.34) suitable either for the construction of helices [201] or turns [202–204].

Alternatively, rigidification of the γ-peptide backbone to avoid H-bonds between nearest neighbors can be achieved by the introduction of an $α,β$-unsaturation into the backbone of each γ-amino acid constituent (vinylogous peptides) [208, 209]. Recent *ab-initio* calculations suggested that the $α,β$-unsaturated γ-peptide backbone might support the formation of helices with large 19- and 22-membered H-bonded pseudocycles [221].

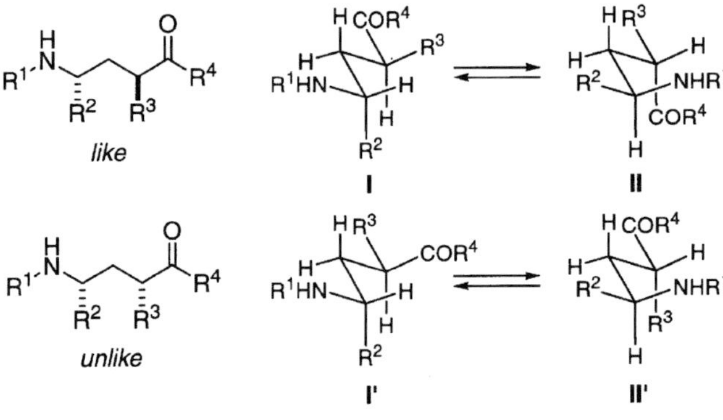

Fig. 2.34 The two conformations free of destabilizing syn-pentane interaction [215, 216] in 2,4-disubstituted γ-amino acid derivatives with *like* and *unlike* configuration. According to the nomenclature proposed by Balaram and by analogy with β-peptides [158], the conformational space of γ-peptides can be described by the following five backbone torsion angles: ω, ϕ, θ_1, θ_2 and ψ

2.3.1
Preparation of γ-Amino Acid Monomers for γ-Peptide Synthesis

As already mentioned γ-amino acids display important biological activities (γ-amino butyric acid (GABA), (R)-carnitine) and are a common structural feature in many natural products and synthetic molecules of pharmaceutical interest. Therefore, a number of approaches have been investigated for their preparation in enantiomerically pure form. Most of them rely on chiral auxiliaries or on the use of the chiral pool (e.g. amino acids, carbohydrates). Although an extensive survey of these methods is beyond the scope of this chapter, the different routes used for the preparation of the various γ-amino acid building blocks with various substitution patterns for γ-peptide synthesis are briefly outlined in this section. Essentially two routes (Scheme 2.8) have been found to be of practical value in the preparation of Boc-protected derivatives **124** from the corresponding α-amino acids: the Wittig–Horner olefination of α-amino aldehydes [200, 201, 222–224] (route 1), and condensation with Meldrum's acid, reduction of the keto functionality and decarboxylative ring closure to pyrrolidinones **125** (route 2) [225].

Overall yields from Boc-α-amino acids are between 40 and 70% depending on the method and the R side-chain [200, 225]. Although feasible, the double Arndt–Eistert homologation of Boc-protected α-amino acids proved unsatisfactory with only low yields of γ-amino acids **124** (<20% over five steps) [200] (for previous studies with Z- and phthalyl- protected α-amino acids, see [226, 227]).

Alkylation of Li-enolates of N-Boc- or TFA-γ^4-amino acid derivatives (e.g. **126**) with various electrophiles has been shown by Hanessian and Schaum [224] to proceed with a high level of 1,3-asymetric induction to yield *unlike*-$\gamma^{2,4}$-amino acid derivatives (e.g. **127**) in good yields (Scheme 2.9). This methodology has been used

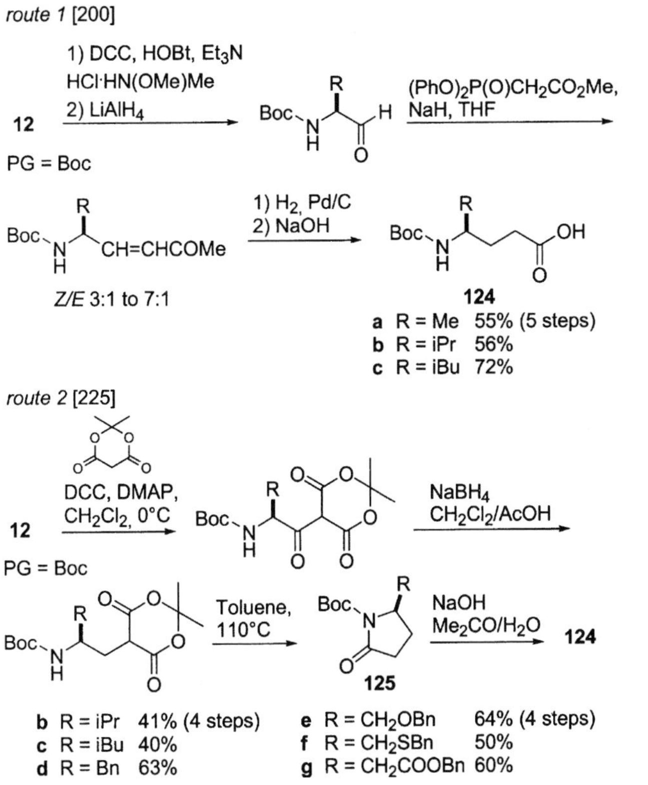

route 1 [200]

1) DCC, HOBt, Et₃N
HCl·HN(OMe)Me
2) LiAlH₄

12

PG = Boc

Boc–NH–CH(R)–CHO

(PhO)₂P(O)CH₂CO₂Me,
NaH, THF

Boc–NH–CH(R)–CH=CHCOMe

Z/E 3:1 to 7:1

1) H₂, Pd/C
2) NaOH

Boc–NH–CH(R)–CH₂CH₂–COOH

124

a R = Me 55% (5 steps)
b R = iPr 56%
c R = iBu 72%

route 2 [225]

12

PG = Boc

DCC, DMAP,
CH₂Cl₂, 0°C

Boc–NH–CH(R)–CH₂–C(=O)–[Meldrum's acid]

NaBH₄
CH₂Cl₂/AcOH

Boc–NH–CH(R)–CH₂–[Meldrum's acid]

Toluene,
110°C

Boc–NH–CH(R)–[pyrrolidinone]

125

NaOH
Me₂CO/H₂O **124**

b R = iPr 41% (4 steps)
c R = iBu 40%
d R = Bn   63%

e R = CH₂OBn   64% (4 steps)
f R = CH₂SBn   50%
g R = CH₂COOBn 60%

Scheme 2.8

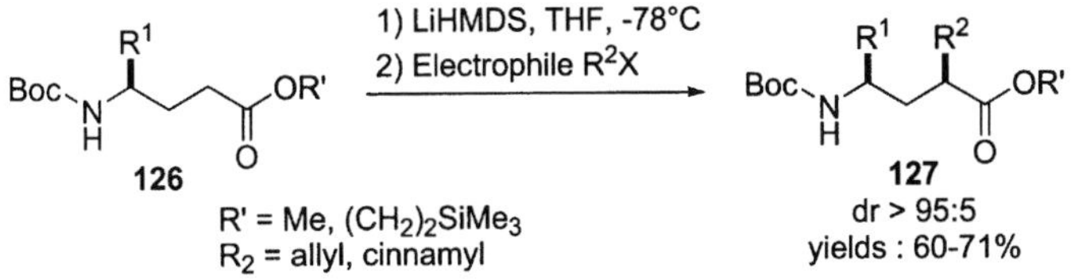

Boc–NH–CH(R¹)–CH₂CH₂–CO–OR'

126

1) LiHMDS, THF, -78°C
2) Electrophile R²X

Boc–NH–CH(R¹)–CH(R²)–CO–OR'

127

dr > 95:5
yields : 60-71%

R' = Me, (CH₂)₂SiMe₃
R₂ = allyl, cinnamyl

Scheme 2.9

to prepare $\gamma^{2,4}$-amino acids **128** with various substitution patterns suitable for the synthesis of turn-inducing γ-peptides [201–204].

Two procedures have been employed for the synthesis of the corresponding *like-*$\gamma^{2,4}$-amino acids **131** and **132** (Scheme 2.10).

α-Methylated derivatives **131** with Me and iPr side-chains at the γ-position were prepared from the *(E)*-α,β-unsaturated γ-amino acid ester **129** [201, 228, 229]. Hydrogenation over Pd/C afforded a 2:1 diasteromeric mixture of *like-* and *unlike* isomers **130** and *epi-***130**. Following saponification and fractional recrystallization,

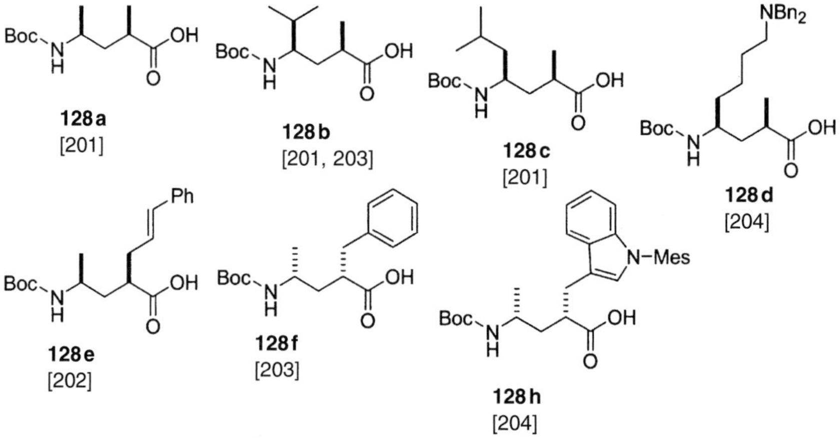

128a
[201]

128b
[201, 203]

128c
[201]

128d
[204]

128e
[202]

128f
[203]

128h
[204]

amino acids **131** were finally obtained in enantiomerically pure form in 39–49% yield from **129**. *like*-$\gamma^{2,4}$-Amino acids **132** were synthesized via stereoselective alkylation of pyrrolidinones **125** in the presence of LiHMDS at −78 °C [202]. The use of amino acids **132** in γ-peptide synthesis was hampered by a marked tendency to lactamization during peptide coupling steps [202].

A general approach to the synthesis of enantiomerically pure γ^2, γ^3 as well as $\gamma^{2,3,4}$-amino acids has been developed by Brenner and Seebach [206, 207, 230]. It involves the Michael addition of Ti-enolates generated from acyl-oxazolidin-2-ones to nitroolefins in the presence of a Lewis acid (TiCl$_4$, Et$_2$AlCl) as the key step

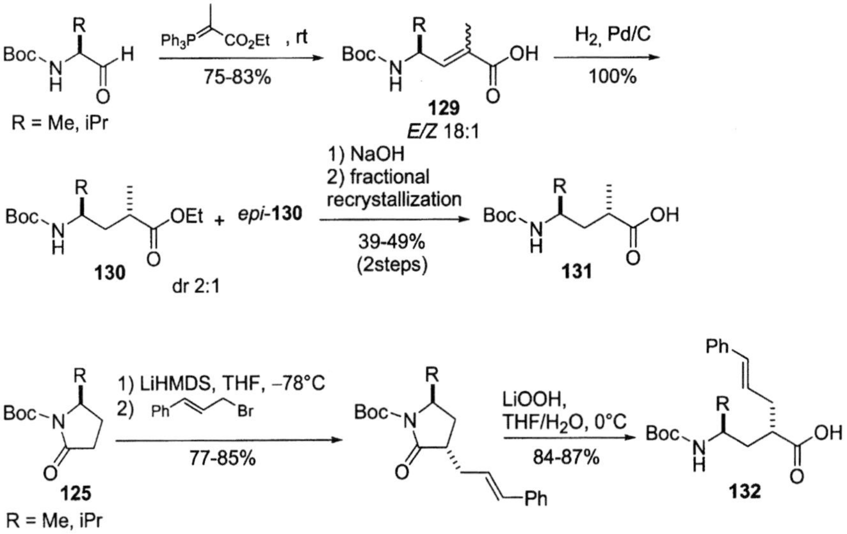

Scheme 2.10

(Scheme 2.11). The reaction allows for the formation of up to three stereogenic centers when 1,2-disubstituted nitroethene derivatives are employed. The resulting γ-nitro carboxylic-acid derivatives **133** and **136** are produced in high diastereoselectivities and are recovered in moderate to good yields. Hydrogenation of the NO_2 group over Raney-Ni results in the formation of the corresponding pyrrolidinones **134** and **137** with concomitant release of the chiral auxiliary. In the case of **137** and **134**, the reaction was accompanied by some epimerization at the α-carbonyl position. Transformation of **134** to the Boc-pyrrolidinone and hydrolysis with LiOH afforded the expected N-Boc-γ²- and γ³-amino acids **135a–c** and **135d–f**, respectively in good yields. Significant racemization frequently occurred at the 2-position during this reaction sequence and γ-amino acids were generally recovered in enantiomerically pure form by crystallization. Treatment of pyrrolidinones **137** with Boc_2O and DMAP did not give the expected Boc-protected derivatives. However, Boc-protected γ²,³,⁴-amino acids **138** could be obtained in moderate yields following direct hydrolysis of **137** by refluxing in 6 M HCl and subsequent N-Boc protection. However, the use of γ²,³,⁴-amino acids **138** in γ-peptide synthesis was

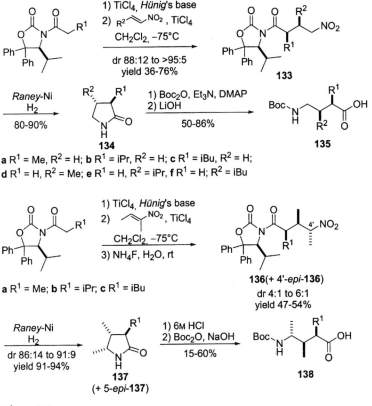

Scheme 2.11

found to be more problematic than use of mono-substituted amino-acids **124** and **135**, due to difficulties encountered in the coupling steps. The use of HATU or EDC/DMAP as coupling agents gave better results than EDC/HOBt.

2.3.2
Helical Folds

Detailed NMR conformational analysis of γ^4-peptides **139–141** (Fig. 2.35) in pyridine-d_5 revealed that γ-peptides as short as four residues adopt a 2.6-helical fold stabilized by H-bonds between C=O$_i$ and NH$_{i+3}$ which close 14-membered pseudocycles [200, 201]. The 2.6_{14}-helical structure of a low energy conformer of γ-hexapeptide **141** as determined from NMR measurements in pyridine-d_5 [200], is shown in Fig. 2.36 A and B). Determination of the structure of γ^4-peptides in CD$_3$OH was hampered by the much lower dispersion of the diasterotopic H-C(α) protons compared to their dispersion in pyridine-d_5. However, the characteristic and properly resolved $i/i+2$ NOE crosspeaks between H-C(γ)$_i$ and NH$_{i+2}$ in the NH/H-C(γ) region of the ROESY spectrum were an indication that the 2.6-helical structure is at least partially populated in CD$_3$OH.

Peptides built from γ-amino acids with L-amino acid-derived chirality centers form a right-handed (P)-2.6_{14} helix of ca. 5 Å pitch with both ethane bonds in a $(+)$-*synclinal* conformation (mean values for dihedral angles θ_1 and θ_2 of central residues 2–5 in compounds **141** are 72.5 and 64.3 °, respectively; Fig. 2.36 A and B; Tab. 2.8).

While the α-helix of L-α-peptides and the (M)-3_{14} helix of the corresponding β^3-peptides have opposite polarity and helicity (see Section 2.2.3.1), the insertion of two CH$_2$ groups in the backbone of L-α-amino acids leave these two helix parameters unchanged, both the α-helix and the 2.6_{14}-helix of the resulting γ^4-peptides being right-handed and polarized from N to C terminus. In view of these similarities, the γ-peptide helical fold might prove useful as a template to elaborate functional mimetics of bioactive α-polypeptides.

The helix dipole in the 2.6_{14}-helical structure with its positive pole at the N-terminus suggest a trend similar to that observed in α-helical α-peptides, that is: (i) the free NH$_3^+$ terminus in unprotected γ-peptides is helix-destabilizing; (ii) negatively-charged residues close to the N-terminus and positively charged residues in close proximity to the C-terminus may exert a stabilizing effect; (iii) removal of the positive charge at the N-terminus via capping (ROCO-, RCO-) may increase

---

Fig. 2.35 γ-Peptides studied by NMR and shown to adopt a 2.6_{14}-helical secondary structure. These include γ-peptides with homochiral sequences consisting of enantiopure γ^4-amino acid residues (**139–142**), $\gamma^{2,4}$-amino acid residues of relative configuration *like* (*l*) (**143–145**) and $\gamma^{2,3,4}$-amino acid residues of relative configurations *like,like* (*l,l*) (**146, 147**). In the case of peptide **146**, the helical structure was also observed in the solid state

γ^4-peptides

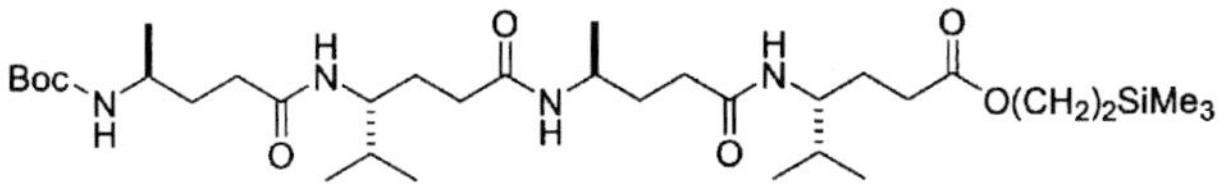

139 [201]

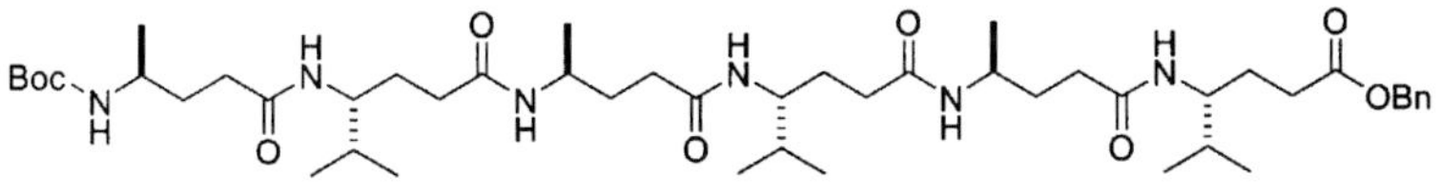

140 [201]

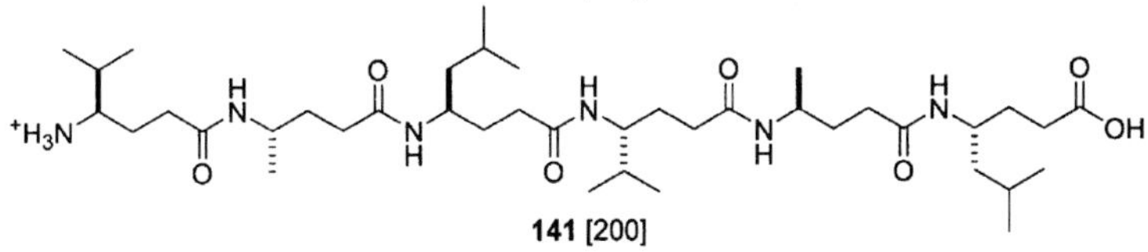

141 [200]

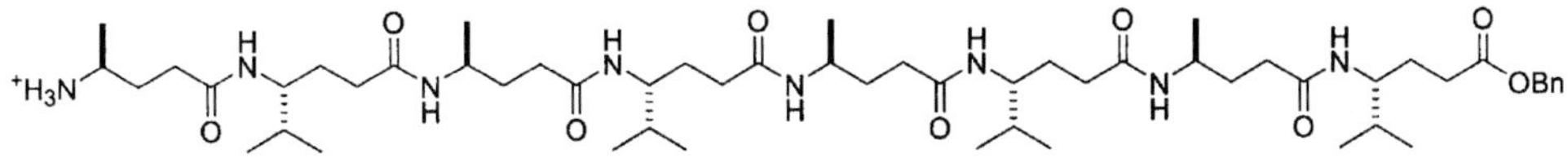

142 [201]

$\gamma^{2,4}$-peptides with residues of relative configuration *like (l)*

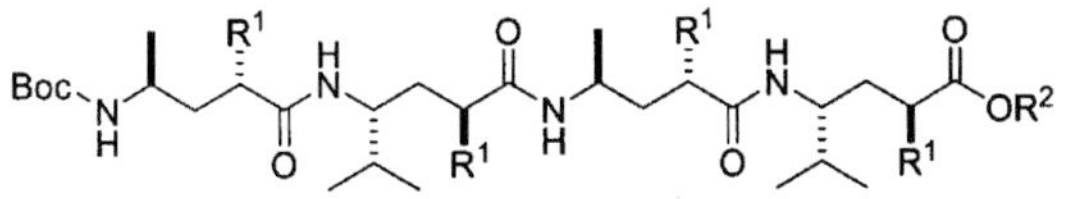

143 R^1 = Me, R^2 = Bn; [201]
144 R^1 = cinnamyl, R^2 = Me; [202]

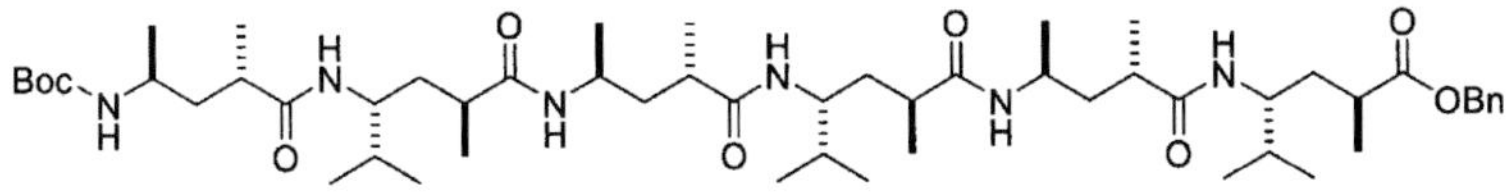

145 [201]

$\gamma^{2,3,4}$-peptides with residues of relative configuration *like,like (l,l)*

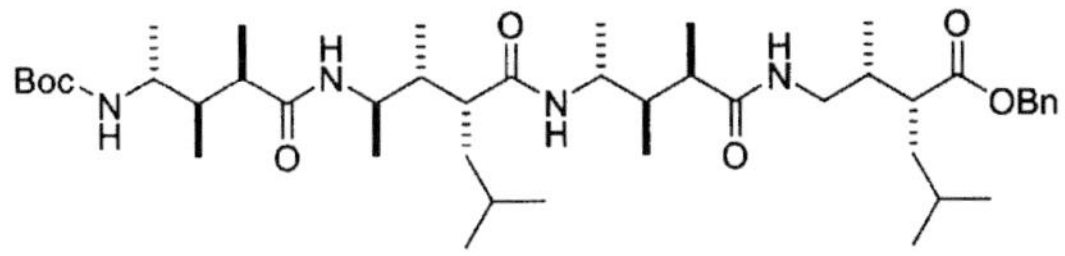

146 [206,207]

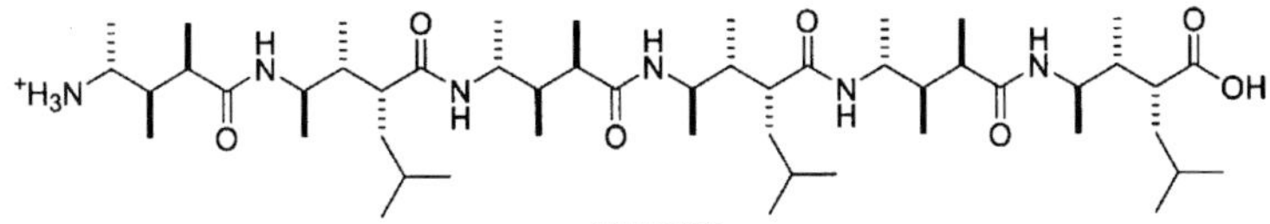

147 [207]

A   B

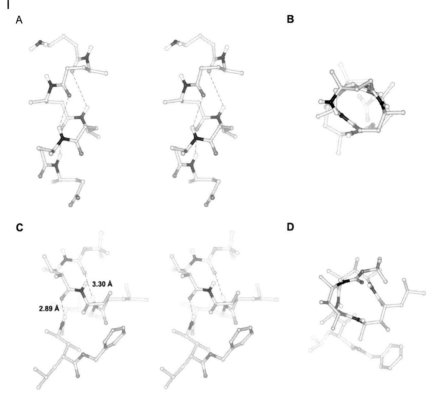

C   D

Fig. 2.36 The γ-peptide 2.6$_{14}$-helical fold. (**A**) Stereo-view along the helix axis of the (*P*)-2.6$_{14}$-helical structure adopted by γ^4-hexapeptide **141** in pyridine. This low energy conformer was obtained by simulated annealing calculations under NMR restraints. Side-chains have been partially omitted for clarity. (Adapted from [200]). (**B**) Top view of **141** (derived from NMR restraints) [200]. (**C**) X-ray crystal structure of γ^{2,3,4}-peptide **146** built with (*R*,*R*,*R*)-amino acids **138a** and **138c** [206, 207]. It is characterized by two H-bonded 14-membered pseudocycles. H-bond N···O distances are shown. Angles (N-H···O) are 171.9 and 154.4°. For the first three residues, both ethane bonds adopt a (–)-synclinal arrangement similar to that found in the (*M*)-2.6-helical backbone. While the side-chain at C(β) forms an angle of approximately 35° with the helix axis, the side-chains at C(α) and C(γ) occupy lateral positions. In contrast, the C-terminal residue is characterized by an antiperiplanar conformation around both ethane bonds ((θ$_1$ value 174.6°, θ$_2$ value –178.6°). (**D**) Top view of **146** (X-ray) [206, 207]

helix stability. This stabilization effect can be due to both the suppression of an unfavorable electrostatic interaction and to the formation of an extra H-bond involving the added H-bond acceptor.

Out of the two conformations of the *like*-γ^{2,4}-amino acid backbone (Fig. 2.34) that do not suffer from *syn*-pentane interaction, conformation **II** is almost identical to that found into the 2.6-helical structure reported for γ^4-peptides. This suggest that avoidance of unfavorable *syn*-pentane interactions in γ-amino acids substituted at both the 2 and 4 positions can be used as an additional element of de-

Tab. 2.8 Comparison of selected backbone torsion angles characteristic of the γ-peptide 2.6-helical backbone extracted from NMR solution structure of γ^4-hexapeptide **141** and solid-state structure of $\gamma^{2,3,4}$-tetrapeptide **146** [200, 206, 207]

Torsion angles	γ-Peptides						
	141[a]				**146**[b]		
	Residue 2	*Residue 3*	*Residue 4*	*Residue 5*	*Residue 1*	*Residue 2*	*Residue 3*
ϕ (°)	−117.5	−121.5	−134.9	−134.1	114.2	150.6	156.4
θ_1 (°)	73.2	71.1	75.5	70.3	−68.7	−71.0	−61.7
θ_2 (°)	65.9	71.1	56.7	63.4	−52.4	−61.1	−50.5
ψ (°)	−139.7	−138.4	−142.5	−140.5	151.1	126.5	123.3

a) From a low energy conformer of γ-peptide **141** generated by molecular modeling under NMR-restraints [200].
b) X-ray crystal structure [206].

sign to enforce optimal pre-organization of the backbone for 2.6-helix formation. The experimental evidence came from studies of γ-peptides **143–147**, consisting either of *like*-$\gamma^{2,4}$ amino acid residues [201, 202] (a cautionary note: in the case of residues bearing a valine side-chain, the CIP priority of substitutents is reversed, thus causing a change in the configurational designation. Hence, $\gamma^{2,4}$-Val(αMe) amino acid residues in peptide **143** are formally of *unlike*-configuration!) or of $\gamma^{2,3,4}$-amino acid residues with relative configuration *l,l* [206, 207] which were all found to adopt a stable 2.6-helical conformation in solution (pyridine-d_5 or CD_3OH). Furthermore, X-ray studies of tetrapeptide **146** revealed the 2.6_{14}-helical motif (first three residues) in the solid state (Fig. 2.36 C and D) [206, 207]. Temperature-dependent NMR studies were undertaken to gain insight into the stability of the γ-peptide helix in solution [201, 202, 207]. Information about the solvent accessibility of NH protons in γ-peptides and their possible engagement in H-bonds can be obtained from temperature coefficients. It is generally assumed [231] that for α-peptides in a rigidly folded state, amide protons with temperature coefficients less negative than −4 ppb K^{-1} are involved in H-bonding. The NH of the first two residues in γ-peptides **139** and **140** display temperature coefficients with particularly high absolute values (15.1 ppb K$^{-1} \geq -\Delta\delta/\Delta T \geq$ 12.6 ppb K^{-1}) in pyridine, thus indicating high solvent accessibility. In contrast, NH of residues 3 and 4 (tetrapeptide **139**) and of residues 3–6 (hexapeptide **140**) exhibit lower absolute values (5.8 ppb K$^{-1} \geq -\Delta\delta/\Delta T \geq$ 3.9 ppb K^{-1}) thus suggesting increased solvent shielding and intramolecularly hydrogen-bonded protons. Interestingly, the same trend is observed for $\gamma^{2,4}$ peptides **143–145** but the absolute values of the temperature coefficients (12.3 ppb K$^{-1} \geq -\Delta\delta/\Delta T \geq$ 9.1 ppb K^{-1} for NH of residues 1 and 2; 3.1 ppb K$^{-1} \geq -\Delta\delta/\Delta T \geq$ 1.7 ppb K^{-1} for the NH groups of the other residues) are consistently and significantly lower than those in the corresponding γ^4-peptides, thus supporting a higher order of helix stability. Additionally, the large ^3J(NH, H-

C(β)) values (characteristic of a nearly antiperiplanar arrangement between these two protons) extracted from ^{1}H-NMR spectra of **147** recorded in CD$_3$OH, decrease only slowly with increasing temperature up to 363 K, thus suggesting that the 2.6$_{14}$-helical structure is maintained to a large extent even at high temperatures. Altogether, these data provide compelling evidence for the stability of the 2.6$_{14}$-helix in solution (pyridine and CD$_3$OH).

Surprisingly, in contrast to α- and β-peptides, CD spectra of γ-peptides gave only a very limited amount of structural information. Experiments conducted on helical γ^4-hexapeptides did not reveal any characteristic CD signature (no Cotton effect) [200, 201]. Similarly, γ^{2,4}-peptides built from 2,4-disubstituted γ-amino acids of *like* configuration and shown to adopt a more stable 2.6-helical structure, do not display typical CD curves either [201]. However, CD spectra of the 2.6-helical γ^{2,3,4}-peptide **147** and its Boc-protected derivative recorded in MeOH and CD$_3$CN present an intense maximum around 215 nm with a shoulder at ca. 200 nm [207].

Structural characterization of γ^3 and γ^2 peptides **148** and **149** by NMR in both pyridine-d_5 and CD$_3$OH was hampered by the low dispersion and strong signal overlap, and so far the conformational preferences of these peptides, if any, remain undetermined.

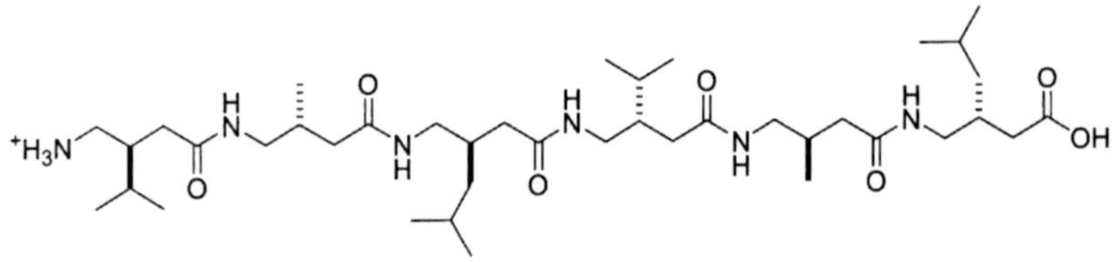

148 [207]

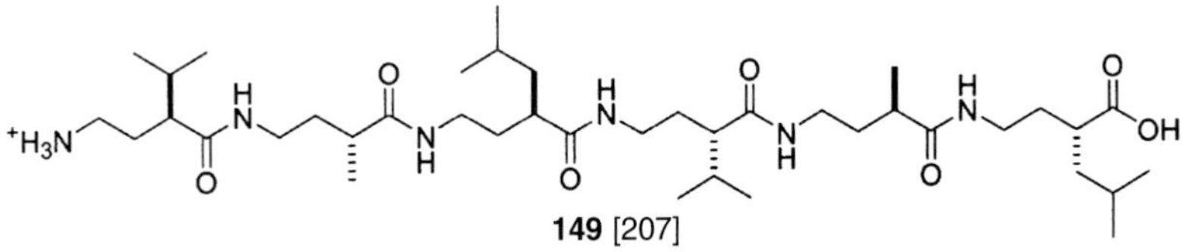

149 [207]

2.3.3
Turn and Sheet Structures

Optimal pre-organization of the γ-peptide backbone towards the formation of open-chain turn-like motifs is promoted by *unlike*-γ^{2,4}-amino acid residues. This design principle can be rationalized by examination of the two conformers free of *syn*-pentane interaction (**I′** and **II″**, Fig. 2.34). Tetrapeptide **150** built from homochiral *unlike*-γ^{2,4}-amino acid building blocks **128e** has been shown by NMR experiments in pyridine to adopt a reverse turn-like structure stabilized by a 14-membered H-bond pseudocycle [202] (Fig. 2.37 A).

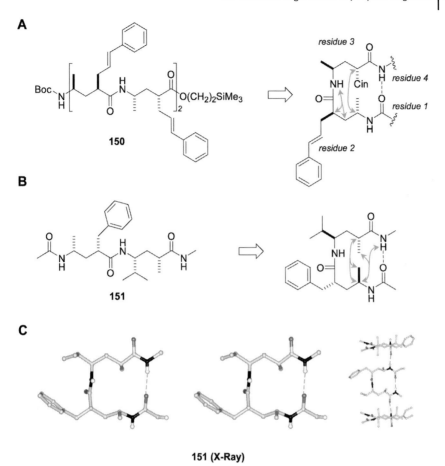

Fig. 2.37 Turn-like motifs induced by γ-peptides consisting of *unlike-γ²·⁴*-amino acid residues. (**A**) 14-membered H-bonded turn segment formed by γ-tetrapeptide **150** with homochiral residues (Cin=cinnamyl). Summary of characteristic NOEs (as arrows) in the turn segment collected in pyridine-*d₅* [202]. (**B**) 14-membered H-bonded turn form by γ-dipeptide **151** consisting of residues of opposite chirality. Summary of NOEs (as arrows) consistent with the turn conformation [203]. (**C**) Structure of **151** in the solid state (stereo-view and detail of the crystal packing). The intramolecular H-bond is characterized by a N···O distance of 2.97 Å and an angle (N-H···O) of 175.5° [203]

The formation of an intramolecular H-bond is supported by the slower rate of amide proton exchange in pyridine/10% CD₃OD. The influence of the stereochemistry on turn formation and turn geometry has been investigated. Seebach and coworkers have demonstrated that *unlike-γ²·⁴* dipeptide sequences consisting of residues with opposite chirality (e.g. **151**) also form a 14-membered H-bonded turn in both solution (CD₃OH) and solid state (Fig. 2.37 B and C). Particularly informative regarding the presence of a turn structure, the NOE connectivity between H-C(γ)ᵢ and H(a)ᵢ₊₁ is observed in the two peptides (*strong* in **150** and *medi-*

um in **151**; see summary of representative NOEs in Figs. 2.37). Interestingly both turns compare well with the type II′ β-hairpin motif of naturally-occurring α-peptides, thus suggesting that short-chain β- and γ-peptides with the correct substitution pattern could be useful as peptidomimetics in drug design (see also Section 2.2.4, Fig. 2.31).

While conformation **II** (Fig. 2.34) of *like-γ*^{2,4}-amino acids is found in the 2.6$_{14}$-helical structure, conformation **I**, which similarly does not suffer from *syn*-pentane interaction, should be an appropriate alternative for the construction of sheet-like structures. However, sheet-like arrangement have not been reported so far for γ-peptides composed of acyclic γ^{2,4}-amino acid residues. Nevertheless, other conformational biases (such as α,β-unsaturation, cyclization between C(α) and C(γ)) have been introduced into the γ-amino acid backbone to restrict rotation around ethylene bonds and to promote extended conformation with formation of sheets in model peptides. Examples of such short chain γ-peptides forming antiparallel (e.g. **152** [208]) and parallel (e.g. **153–155** [205, 208]) sheet-like structures are shown in Fig. 2.38.

The backbone conformation in α-substituted vinylogous dipeptide **152** derives in part from minimization of allylic A(1,3) strain which causes H-C(γ) to lie in the plane of the enamide ($\theta_1 = 134°$). The resulting extended conformation leads to the formation of a two-stranded antiparallel sheet (Fig. 2.38 A). Higher ordered structures are not formed in such peptides presumably because of the steric hindrance generated by the α-methyl. In the absence of allylic A(1,3) strain, the extended conformation which is maintained ($\theta_1 = 116.5°$) in the corresponding dipeptide **153**, gives rise to an infinite parallel β-sheet arrangement (Fig. 2.38 B). However longer peptides have not been studied.

Alternatively, *trans*-3-aminocyclopentanecarboxylic acid (*trans*-3-ACPC), a conformationally constrained cyclic γ-amino acid in which the C(α), C(β) and C(γ) atoms are part of a five-membered ring (i.e. the "γ-version" of the *trans*-2-aminocyclopentanecarboxylic acid, see Section 2.2.3.3.), has been found to be suitable for the construction of a γ-peptide sheet structure [205]. Similar to β-peptide **123** (Section 2.2.4.), γ-peptide segments composed of *trans*-3-ACPC residues connected by the D-prolyl-(1,1-dimethyl)-1,2-diaminoethyl (Pro-DADME) unit (e.g. **154**, **155**), form a parallel hairpin in both solution and solid state (Fig. 2.38 C). NMR analysis of **154** in CD$_2$Cl$_2$ at a concentration where aggregation does not occur, revealed that the signals of the two NH protons on the H-bonded donor strand are strongly shifted downfield thus supporting the existence of substantial intramolecular H-bonding. In addition the two peptides display a characteristic NOE pattern (see summary of NOEs in Fig. 2.38 C) consisiting of (i) a short range NOE between H-C(δ) of the proline ring and H-C(α) of the adjacent *trans*-3-ACPC residue which supports a *trans*-conformation of the tertiary amide bond, and (ii) strong interstrand backbone NOE connectivities between H-C(γ) on the H-bonded acceptor strand and H-C(α) on the H-bonded donor strand.

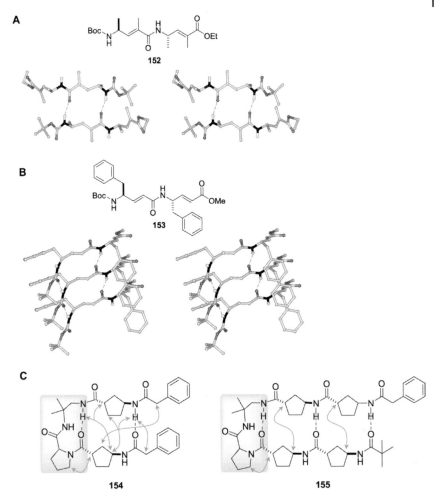

Fig. 2.38 Sheet forming γ-peptides. (**A**) Crystal structure of the two stranded antiparallel sheet formed by α,β-unsaturated γ-dipeptide **152** with α-methyl substituted residues [208]. Both intermolecular H-bonds are characterized by a N···O distance of 2.84 Å and an angle (N-H···O) of 164.2°. (**B**) Crystal structure of the infinite parallel sheet arrangement formed by vinylogous dipeptide **153** [208]. Intermolecular H-bonds are characterized by a N···O distance of 2.88 Å and 3.24 Å and an angle (N-H···O) of 160.3 and 167.0°, respectively. (**C**) Structure of parallel hairpins **154** (solution and solid state) and **155** (solution). The two extended γ-peptide chains composed of trans-3-ACPC residues are connected by a D-prolyl-(1,1-dimethyl)-1,2-diaminoethyl (Pro-DADME) unit. Characteristic NOEs (as arrows) collected in CD_2Cl_2 (3.6 mM for **154**) and pyridine-d_5 (2 mM for **155**) are summarized [205]

2.4
Biological Activities of β- and γ-Peptides

2.4.1
Biological Stability

The capacity of β- and γ-peptides to adopt well-defined and controlled secondary structures including helices, sheet-like and turn structures depending on their substitution pattern and stereochemistry (see previous sections), suggests that such rigid backbones can be used as scaffolds to place and orient pharmacophores in a predictable manner for *de novo* design of molecules with interesting biological activities. The schematic representations of idealized β- and γ-peptide helices looking down the helix axis (helical wheel diagram) as well as their comparison with the α-helical wheel (Fig. 2.39) are particularly informative when elaborating the functional mimetics of α-helical surfaces.

In contrast to natural α-polypeptides, β- and γ-peptides display remarkable *in vitro* stability to degradation by peptidases from bacterial, fungal and eukaryotic origins (e.g. leucyl aminopeptidase, proteinase K, trypsin, elastase, amidase, β-lactamase, 20 S proteasome, etc.) which makes them even more attractive for biomedical applications [10, 232–234]. Bioavailability is another important issue for the possible use of β- and γ-peptides in drug design. This point has recently been addressed in the case of the β-nonapeptide H-(β^3-HAla-β^3-HLys-β^3-HPhe)$_3$-OH (156a) [235]. Pharmacokinetic studies on this peptide, ^{14}C-labeled at C=O and C(a) of residue 6, revealed in rats that following i.v. administration of 5 mg kg^{-1}: i) the concentration of β-peptide in blood and plasma decreases rapidly and the radioactivity is distributed in the organs and tissues; (ii) after 1 week, residual radioactivity in organs and tissues represented 49% of the i.v. dose with high levels in the kidney, lymph nodes and liver; (iii) the β-peptide is highly stable against metabolic degradation *in vivo*. Negligible absorption took place after p.o. administration, and the administered dose was completely excreted via the feces within 96 h [235]. Finally, β-amino acids that may eventually result from β-peptide degradation were found to be non-mutagenic according to Aimes' test [232].

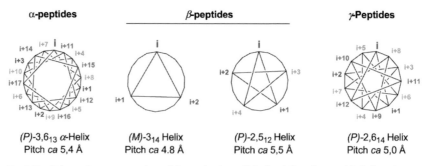

α-peptides	β-peptides		γ-Peptides
(P)-3,6$_{13}$ α-Helix Pitch *ca* 5,4 Å	(M)-3$_{14}$ Helix Pitch *ca* 4.8 Å	(P)-2,5$_{12}$ Helix Pitch *ca* 5,5 Å	(P)-2,6$_{14}$ Helix Pitch *ca* 5,0 Å

Fig. 2.39 Schematic representation of the projection of idealized β- and γ-peptide helices in a plane perpendicular to the helix axis and comparison with the helical wheel of the natural α-helix

2.4.2
Bioactive Peptides Based on Helical Scaffolds

Research activity in this area has mainly concentrated on the design and *in vitro* studies of amphiphilic helical β-peptides with antimicrobial activity. In view of their high propensity for helical conformations as well as their resistance to proteolytic degradation, β-peptides represent promising antibacterial candidates.

Antimicrobial α-peptides that adopt cationic amphipatic α-helical structures upon binding to cell membranes (e.g. mellitin from bee venom [236], magainins from frog skin [237], cecropins from porcine small intestine [238]) are ubiquitous in nature (Antimicrobial Sequences Database, *http://www.bbcm.units.it/~tossi/pag1.htm*) and represent important effectors molecules of innate immunity (for reviews, see [239, 240]). These peptides generally cause cell death by a two-step mechanism involving interaction with the lipid component of the membrane (while in bacteria the outer leaflet of the membrane is essentially composed of lipids with negatively-charged phospholipid headgroups, it is almost neutral in plants and animals) followed by membrane permeabilization. Several mechanisms for membrane permeabilization have been postulated including transient pore formation ("barrel-stave" model) or detergent-like diruption of the membrane ("carpet" model) [241, 242]. The lytic activity of amphiphilic antimicrobial peptides is strongly influenced by properties such as helix stability, amphiphilicity (hydrophobic moment), hydrophobicity, relative width of the hydrophilic and hydrophobic faces of the helix as well as net charge [243]. Despite many structure–activity studies, methods of optimization remain challenging because sequence modifications of α-peptides generally affect several parameters at the same time. In addition, low activity on human cell membranes is a prerequisite for a cell-lytic peptide to be of therapeutic value and *de-novo* design of helicoidal membrane-lytic peptides with high membrane selectivity necessitates even finer tuning between these five different parameters.

Both the 3_{14}- and 2.5_{12}-helical backbones have been found suitable for the design of antimicrobial β-peptides. In order to cluster polar residues on one face of the helix, amphiphilic 3_{14}-helical β-peptides have been constructed from hydrophobic–cationic–hydrophobic or hydrophobic–hydrophobic–cationic residue triads

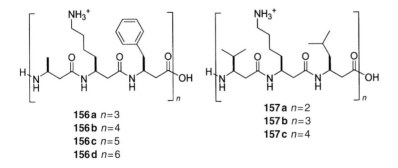

156a *n*=3
156b *n*=4
156c *n*=5
156d *n*=6

157a *n*=2
157b *n*=3
157c *n*=4

[175, 244–246]. Early studies by DeGrado [244, 245] and Seebach [246] focused on β^3-peptides of various lengths (6–18 residues) with a free *C*-terminal carboxylic acid and assembled from β^3-HAla-β^3-HLys-β^3-HPhe (e.g. **156a–d**), β^3-HVal-β^3-HLys-β^3-HLeu (e.g. **157a–c**), or β^3-HLeu-β^3-HLys-β^3-HLeu triads.

Four triad repeats (e.g. **156b** and **157c**) was found to be the optimal length to obtain antibacterial activity (albeit modest) against *E. coli* (Gram-negative): the minimum inhibitory concentration (MIC) required for complete inhibition of growth by **156b** [246] and **157c** [245] was 32 μg mL^{-1} (strain NB 27001) and 9 μg mL^{-1} (strain K91), respectively. However, with a concentration of 37 μM required to lyse 50% of human erythrocytes, peptide **157c** was shown to be strongly hemolytic. As previously discussed in Section 2.3.1., these β^3-peptides show essentially no helical content in aqueous solution, but in the presence of DPC micelles [244] or phospholipid membranes [245], their CD spectra are characteristic of a large population of 3_{14}-helices. In a structure-activity study aimed at probing the relationship between helix stability and antibacterial activity, Gellman found that the replacement of the *C*-terminal carboxylic acid of nonapeptide *ent*-**157b** by a carboxamide resulted in a dramatic increase in the antibacterial potency (see Tab. 2.9) [175].

The resulting β^3-nonapeptide **85** as well as its enantiomer *ent*-**85** were both found to be active against Gram-positive (*S. aureus* and *E. faecium* strains were clinical isolates resistant to penicillin and vancomycin, respectively) and Gram-negative bacteria with MIC values (in the range 1.6–12.5 μg mL^{-1}) comparable to those of the antimicrobial α-peptides melittin and [Ala^{8,13,18}]-magainin II amide (a highly potent synthetic analog of natural magainin II [247]). Although peptide **85** shows no helix formation in water, it displays a maximimum helicity in 40% aqueous TFE, a solvent system which is also known to promote helicity of amphiphilic α-peptides. Interestingly, rigidification of the backbone by insertion of one-third (e.g. **84**) and two-thirds *trans*-ACHC (e.g. **83**) residues to increase helix propensity in aqueous solution (see Fig. 2.15; Section 2.2.3.1) has only little effect on

Tab. 2.9 Antibacterial activities of [Ala^{8,13,18}]-magainin II amide and amphiphilic 3_{14}-helical (*ent*-**157b**, **83**–**85**) and 2.5_{12}-helical (**158**, **159**) β-peptides [175, 234, 248]

Bacterial strain	Peptide								
	[Ala^{8,13,18}]-magainin II amide	*ent*-157b	85	84	83	158a	158b	159a	159b
E. coli JM109	6.3	200	12.5	12.5	6.3	6.3	25	100	>200
B. subtilis BR151	3.1	50	1.6	0.8	0.8	0.8	1.6	50	25
S. aureus 1206	25–50	200	6.3	3.1	6.3	3.2	12.5	>200	>200
E. faecium A634	25	≥200	12.5	3.1	3.1–6.3	12.5	50	>200	>200

Activities are expressed as minimum inhibitory concentration (MIC, μg mL^{-1}) defined as the lowest concentration required for complete inhibition of growth of bacterial strain.

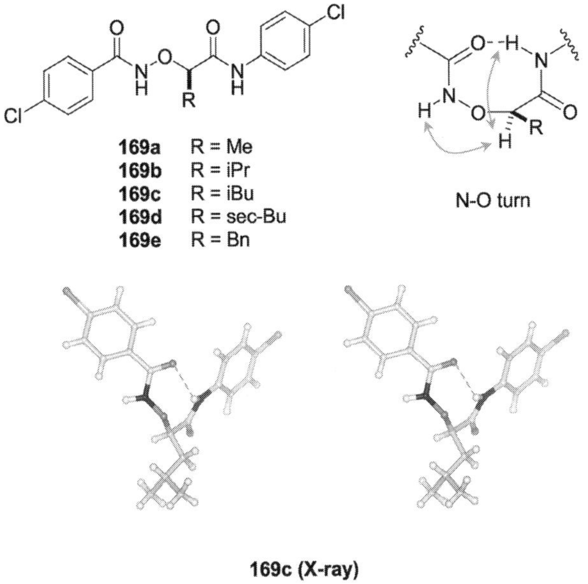

169a	R = Me
169b	R = iPr
169c	R = iBu
169d	R = sec-Bu
169e	R = Bn

N-O turn

169c (X-ray)

Fig. 2.44 The N-O turn formed from *N*-acetylated α-aminooxy carboxamides of D-configuration **169**. Schematic representation of the N-O turn with representative NOEs found in ROESY of **169c** in CDCl₃ and the X-ray crystal structure of **169c**. The intramolecular H-bond is characterized by a short H···O distance of 2.12 Å. According to the nomenclature proposed by Balaram and by analogy with β-peptides [158], the conformational space of oligo(α-aminoxy acid) can be described by the following torsion angles: ω, ϕ, θ_1, and ψ. With a θ_1 value of + 78.5°, the turn is right handed

Interestingly, β-aminoxy acids which are homologs of γ-amino acids have also been found to promote the formation of turns and helices. In apolar solvent and in the solid state, model diamides consisting of $\beta^{2,2}$-aminoxy acid residues adopt a novel N-O turn stabilized by both a nine-membered H-bonded ring between $C=O_i$ and NH_{i+2}, and a six-membered ring formed between $N-O_i$ and NH_{i+1}. The X-ray crystal structure of a corresponding triamide revealed two consecutive C9 N-O turns suggesting a novel 1.7₉-helical fold [279].

2.5.2
Example 2: *N,N'*-Linked Oligoureas as γ-Peptide Mimetics

Enantiopure *N,N'*-linked oligoureas were originally described in 1995 by Burgess and coworkers as novel peptide backbone mimetics [271, 272]. Several synthetic approaches have been reported, all of which involve sequential acylation and amine deprotection cycles using appropriately protected carbonyl synthons [83, 271, 272, 274, 286–288]. Although elongation can be performed in solution, most of the synthetic procedures are elaborated on solid supports starting from Rink's

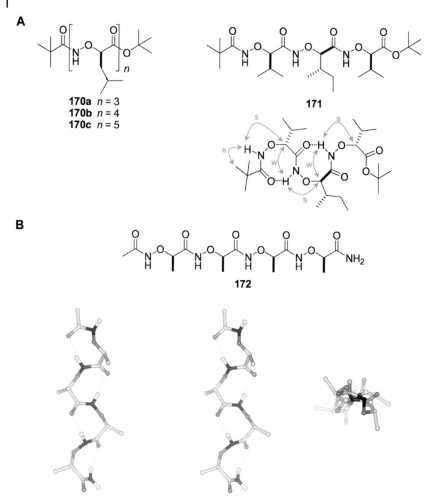

Fig. 2.45 Preferential conformation of α-aminoxy peptide oligomers. **(A)** Formulae of oligo(α-aminoxy acid) **170** and **171** studied by the combination of NMR and CD as well as a summary of key NOEs in **171** observed by ROESY in CDCl₃. **(B)** Theoretical model generated by *ab-initio* quantum-mechanical calcula-tions of the (*P*)-1.8₈-helical fold formed by tetramer **172** (stereo-view along the helix axis and top view). Intermolecular H-bonds are characterized by an angle (N-H···O) of 157°. The side-chains at position *i* and *i*+2 are separated by a distance of 6.5 Å

amide resin [289]. Preparation of activated monomers (i.e. 2-phthalimido-1-substituted-ethylisocyanates **173** [271, 272], 4-nitrophenyl 2-azido-1-substituted-ethyl carbamates **174** [286], 4-nitrophenyl 2-[(tert-butoxycarbonyl)amino]-2-substituted-ethyl carbamates **175** [287]) generally involves reaction of a monoprotected diamine with a carbonylating agent such as phosgene or *p*-nitrophenyl chloroformate. Curtius rearrangement of *N*-protected β-amino acyl azides to generate succinimidyl carbamates **176** (e.g. succinimidyl 2-[(tert-butoxycarbonyl)amino]-2-substituted-ethyl carbamates [83], succinimidyl 2-{[(9H-fluoren-9-ylmethoxy)carbonyl]amino}-2-substituted-ethyl carbamates [274, 288]) has also been reported. The conformational preference and folding propensity of two oligoureas (i.e. heptamer **177** and nonamer **178**) synthesized on a solid support using succinimidyl carbamates **176b** have been investigated by NMR (DQF-COSY, TOCSY, and ROESY) and circular dichroism [273, 274] (for the nomenclature, see Fig. 2.46 A)

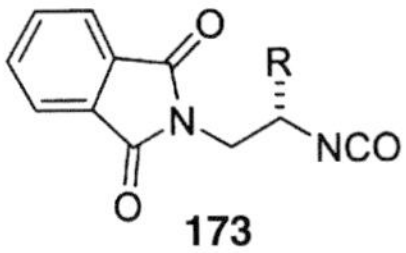

173 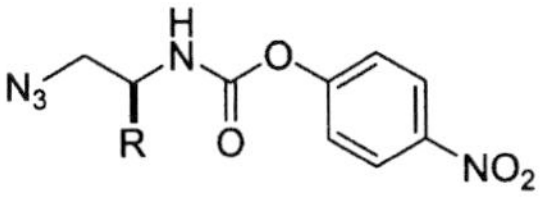　　**174**

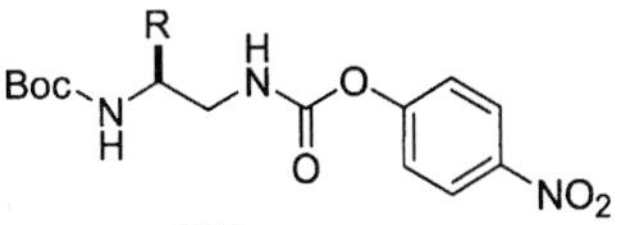

175 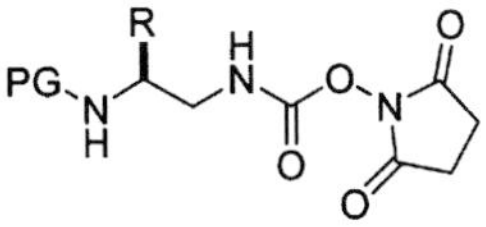　　**176**

a PG = Boc; **b** PG = Fmoc

A number of data extracted from ^{1}H spectra of **177** and **178** recorded in pydidine-d_5 were indicative of conformational homogeneity. These include large $^3J(NH, H-C(\beta))$ values (>9 Hz for central residues), large chemical shift differences between diastereotopic H-C(a) protons (1.0 < $\Delta\delta$ <1.6 ppm for all residues except number one), strong differentiation between vicinal coupling constants of each pair of diastereotopic H-C(a) along the sequence, as well as very little temperature dependence (up to 362 K) for these parameters. The non-equivalence of

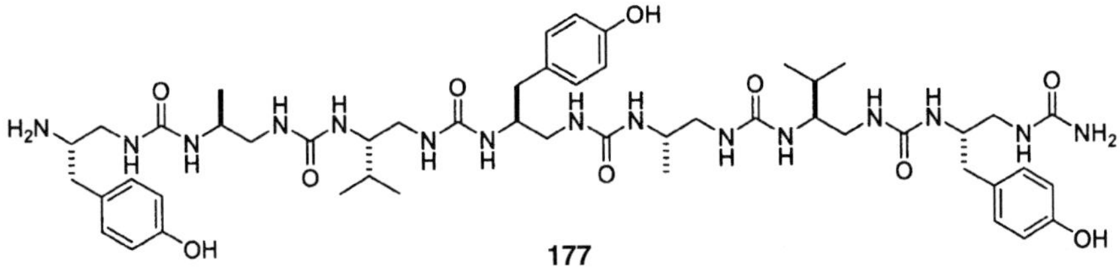

177

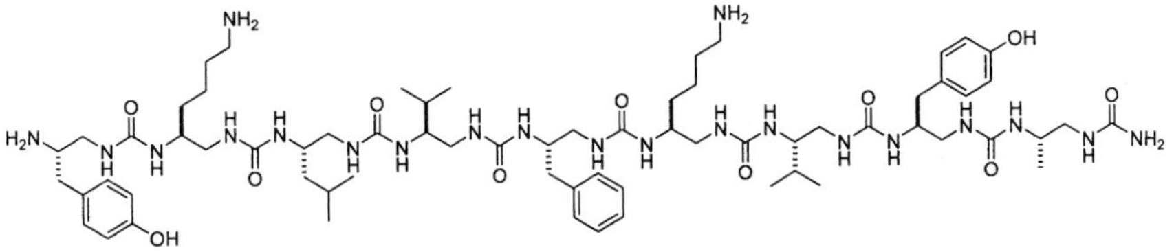

178

diastereotopic $C(a)$ protons for central residues at high temperature was consistent with their location in a distinct spatial environment and was taken as an indication of a well-defined and stable secondary structure. Measurements of ^3J(H-$C(\beta)$, H-$C(a)$) coupling constants and intensities of intraresidue NOE between NH and $C(a)$ protons allowed the unambiguous stereospecific assignment of diastereotopic $C(a)$ protons, the upfield proton being assigned to H^{Si}-$C(a)$, as well as the determination of the conformation around the $C(a)$-$C(\beta)$ bond (i.e. H^{Si}-$C(a)$ is nearly antiperiplanar to NH and synclinal to H-$C(\beta)$) (Fig. 2.46).

Additional information about the three-dimensional structure came from ROESY data. Medium-range $i/i+2$ connectivities of the type H-$C(\beta)_i$/NH$_{i+2}$, H-$C(\beta)_i$/NH$_{i+2}$ and H^{Si}-$C(a)_i$/NH$_{i+2}$ repeated along the sequence in **177** and **178** were consistent with the presence of a helical conformation. This was confirmed by restrained molecular dynamics calculations. In pyridine-d_5, enantiopure N,N'-linked oligoureas **177** and **178** adopt a well-defined 2.5-helical structure, reminiscent of the 2.6$_{14}$-helical structure of γ^4-peptides [273, 274] (Fig. 2.47). As in β- and γ-peptides, the helix of oligoureas is characterized by a stable (+)-*synclinal* arrangement around the ethane bond. However the C=O···H-N intramolecular hydrogen bonding pattern is quite different. In oligoureas, the structure is held by H-bonds closing 12- and 14-membered rings formed between C=O$_i$ and N'H$_{i+2}$ and NH$_{i+3}$, respectively. The side-chains of residues i and $i+5$ are almost superimposed on one another (Fig. 2.47 B). The distance between corresponding H-$C(\beta)_i$ and H-$C(\beta)_{i+5}$ determined from an average structure is between 9.8 and 11.2 Å. With an internal diameter of ca. 3 Å, the helix is particularly compact and is devoid of empty volume in the interior.

Additional information about the solvent accessibility of NH protons in **177** and **178** and their engagement in H-bonds was gathered from temperature coefficients. In the case of the first two urea bonds, the absolute values of the temperature coefficients ≥ 5 ppb K^{-1}, indicated high solvent accessibility. In contrast, NH and N'H of the other residues show temperature coefficients with absolute values < 4 ppb K^{-1}, suggesting solvent shielding and intramolecularly hydrogen-bonded protons.

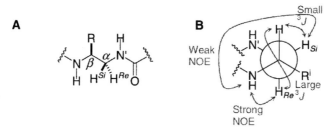

Fig. 2.46 **(A)** Nomenclature for the description of the atoms within each oligourea residue. **(B)** Preferential conformation along the $C(a)$-$C(\beta)$ bond in a Newman projection together with the representative NMR data which validate the model

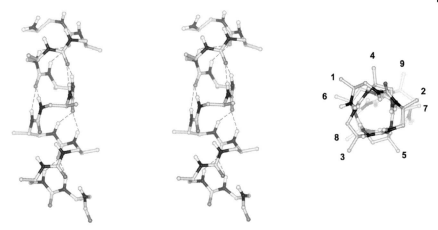

Fig. 2.47 The (*P*)-2.5-helical structure of *N,N'*-linked oligoureas as determined by NMR meaurements in pyridine-d_5. **(A)** Stereo-view along the helix axis of a low energy conformer of nonamer **178** generated by restrained molecular dynamics calculations. (Adapted from [274]). The helix is characterized by (i) a rigid (+)-*synclinal* arrangement around the C(*a*)-

C(*β*) bond very similar to that which is observed for helical *β*- and *γ*-peptides and (ii) two types of intramolecular H-bonds involving a given C=O, namely: C=O$_i$···H–N$'_{i+2}$ and C=O$_i$···H–N$_{i+3}$ which close the 12- and 14-membered H-bonded rings, respectively. **(B)** Top view showing the numbering of the residues

Although the *n*-*π** and *π*-*π** electronic transitions of the urea chromophore have not been studied as extensively as amides, the contribution of the backbone is expected to dominate the far UV spectra of oligoureas in a fashion similar to that which is observed for peptides. The CD spectra recorded in MeOH of oligoureas **177** and **178** show an intense maximum near 204 nm (Figure 2.48). This is in contrast to helical γ^4-peptides that do not exhibit any characteristic CD signature.

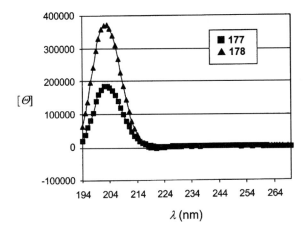

Fig. 2.48 CD analysis of oligoureas **177** and **178**. CD spectra were recorded in MeOH at room temperature at a concentration of 0.2 mM. Molar ellipticity [Θ] in deg cm^2 dmol^{-1}. The mean-residue ellipticities measured at 204 nm for **177** and **178** are 26600 and 41300 deg cm^2 dmol^{-1}, respectively

However and as previously discussed, preferred conformations of novel folding oligomers with chromophores cannot be deduced simply from CD measurements. Complementary NMR studies in the same solvent have thus been conducted to possibly assign the characteristic CD signature to a given conformation. Preliminary results seem to indicate that oligoureas are able to adopt a helical conformation in CD_3OH and that it is indeed possible to correlate their far UV chiroptical properties and their folding propensity (G. Guichard, unpublished results). By analogy to the work on β- and γ-peptides, these conformational studies provide the first step towards the rational design of bioactive oligoureas.

2.6
Conclusion

The β- and γ-peptide backbones display unique properties of self-organization at the molecular level leading to a wealth of stable secondary structures (helical folds, turns, and sheet structures) depending on the substitution pattern and relative configuration of their amino acid constituents. Detailed structural analyses aimed at elucidating the requirements for secondary structure formation have been conducted and the body of structural information now available provides a rationale for the design of foldamers with useful biological activities. In this field, encouraging results have been obtained e.g. amphiphilic helical β-peptides with antimicrobial activity and open-chain β-peptide turn mimics. These artificial systems based on short chain synthetic oligomers designed to fold into regular structures also provide useful models for studying the factors that govern secondary and tertiary structure formation in biopolymers. Finally, similar to α-peptides, the β- and γ-peptide backbones provide a unique source of inspiration for the creation of isosters with a high propensity for novel secondary structures which can mimic those found in proteins.

Note Added in Proof

Seebach and coworkers have prepared seventeen of the twenty β^2-amino acid derivatives with proteinogenic side chains in multigram amounts by diastereoselective reaction of enolates of the corresponding 3-acyl-4-isopropyl-5,5-diphenyloxazolidin-2-ones with appropriate electrophiles [290]. In a short review article, Martinek and Fülöp have highlighted the design principles for side-chain control of β-peptide secondary structures [291]. Seebach and coworkers have shown that β-peptides (with hairpin or helical secondary structure) incorporating β^3-HCys and β^3-HHis residues are capable of binding divalent metal ions such as Zn^{2+} [292]. Kimmerlin et al. have disclosed preliminary results demonstrating the existence of ordered interactions between a helical β-peptide (bearing Asn and Lys side chains) and DNA [293]. Gellman and Diederichsen have designed helical β-peptides incorporating nucleobases that self-assemble through antiparallel base pairing to form stable duplexes [294]. A poly-cationic Tat (47–57) β-peptide analogue has been shown to bind selectively to TAR RNA by Rana and Gellman [295]. Recent results from Gellman and Menon suggest that a polycationic β-dodecapeptide assembled of β^3-HVal-β^3-HArg-β^3-HArg triad enters cells by endocytosis and accumulates in the nucleolus [296].

2.7
References

1 J. Venkatraman, S.C. Shankaramma, P. Balaram, *Chem. Rev.* **2001**, *101*, 3131–3152.

2 W.F. DeGrado, C.M. Summa, V. Pavone, F. Nastri, A. Lombardi, *Annu. Rev. Biochem.* **1999**, *68*, 779–819.

3 R.B. Hill, D.P. Raleigh, A. Lombardi, W.F. DeGrado, *Acc. Chem. Res.* **2000**, *11*, 745–754.

4 S.H. Gellman, *Curr. Opin. Chem. Biol.* **1998**, *2*, 717–725.

5 E. Lacroix, T. Kortemme, M. Lopez de la Paz, L. Serrano, *Curr. Opin. Struct. Biol.* **1999**, *4*, 487–493.

6 Appella, L.A. Christianson, I.L. Karle, D.R. Powell, S.H. Gellman, *J. Am. Chem. Soc.* **1996**, *118*, 13071–13072.

7 S.H. Gellman, *Acc. Chem. Res.* **1998**, *31*, 173–180.

8 D.J. Hill, M.J. Mio, R.B. Prince, T.S. Hughes, J.S. Moore, *Chem. Rev.* **2001**, *101*, 3893–4011.

9 J.A. Patch, A.E. Barron, *Curr. Opin. Chem. Biol.* **2002**, *6*, 872–877.

10 D. Seebach, M. Overhand, F.N.M. Kühnle, B. Martinoni, L. Oberer, U. Hommel, H. Widmer, *Helv. Chim. Acta* **1996**, *79*, 913–941.

11 B. Iverson, *Nature* **1997**, *385*, 113–115.

12 S. Borman, *Chem. Ing. News* **1997**, *June 16*, 32–35.

13 U. Koert, *Angew. Chem. Int. Ed.* **1997**, *36*, 1836–1837.

14 E. Martuscelli, R. Gallo, G. Paiaro, *Makromol. Chem.* **1967**, *103*, 295–298.

15 J.D. Glyckson, J. Applequist, *J. Am. Chem. Soc.* **1971**, *93*, 3276–3281.

16 H. Bestian, *Angew. Chem. Int. Ed. Engl.* **1968**, *7*, 278–285.

17 J. Sebenda, I. Hauner, J. Biros, *J. Polym. Sci. Polym. Chem. Ed.* **1976**, *14*, 2357.

18 F. Chen, G. Lepore, M. Goodman, *Macromolecules* **1974**, *7*, 779–783.

19 C.D. Eisenbach, R.W. Lenz, M. Duval, R.H. Marchessault, *Makromol. Chem.* **1979**, *180*, 429–440.

20 J. Kovacs, R. Ballina, R.L. Rodin, D. Balasubramanian, J. Appelquist, *J. Am. Chem. Soc.* **1965**, *87*, 119–120.

21 P.M. Hardy, J.C. Haylock, H.N. Rydon, *J. Chem. Soc. Perkin Trans. 1*, **1972**, 605–611.

22 H. Yuki, Y. Okamoto, Y. Taketani, T. Tsubota, Y. Marubayashi, *J. Polym. Sci., Polym. Chem. Ed.* **1978**, *16*, 2237–2251.

23 H. Yuki, Y. Okamoto, Y. Doi, *J. Polym. Sci., Polym. Chem. Ed.* **1979**, *17*, 1911–1921.

24 J.M. Fernández-Santín, J. Ayamí, A. Rodríguez-Galán, S. Muoz-Guerra, J.A. Subirana, *Nature* **1984**, *311*, 53–54.

25 J.M. Fernández-Santín, S. Muoz-Guerra, A. Rodríguez-Galán, J. Ayamí, J. Lloveras, J.A. Subirana, E. Giralt, M. Ptak, *Macromolecules* **1987**, *20*, 62–68.

26 J. Bella, C. Alemán, J.M. Fernández-Santín, S.C. Alegre, J.A. Subirana, *Macromolecules* **1992**, *25*, 5225–5230.

27 C. Alemán, J. Bella, J.J. Perez, *Polymer*, **1994**, *35*, 2596–2599.

28 J.J. Navas, C. Alemán, F. López-Carrasquero, S. Muoz-Guerra, *Macromolecules*, **1995**, *28*, 4487–4494.

29 C. Alemán, J.J. Navas, S. Muoz-Guerra, *J. Phys. Chem.* **1995**, *99*, 17653–17661.

30 J.J. Navas, C. Alemán, S. Muoz-Guerra, *J. Org. Chem.* **1996**, *61*, 6849–6855.

31 F. López-Carrasquero, M. Garcia-Alvarez, J.J. Navas, C. Alemán, S. Muoz-Guerra, *Macromolecules* **1996**, *29*, 8449–8459.

32 J.J. Navas, C. Alemán, F. López-Carrasquero, S. Muoz-Guerra, *Polymer* **1997**, *38*, 3477–3484.

33 S. León, C. Alemán, S. Muoz-Guerra, *Macromolecules* **1997**, *30*, 6662–6667.

34 C. ALEMÁN, J. J. NAVAS, S. MUOZ-GUERRA, *Biopolymers*, **1997**, *41*, 721–729.

35 M. GARCIA-ALVAREZ, S. LEÓN, C. ALEMÁN, J. L. CAMPOS, S. MUOZ-GUERRA, *Macromolecules*, **1998**, *31*, 124–134.

36 K. A. BODE, J. APPLEQUIST, *Macromolecules* **1997**, *30*, 2144–2150.

37 H. M. MÜLLER, D. SEEBACH, *Angew. Chem. Int. Ed. Engl.* **1993**, *32*, 477–502.

38 D. SEEBACH, M. G. FRITZ, *Biol. Macromol.* **1999**, *25*, 217–236.

39 S. DAS, U. D. LENGWEILER, D. SEEBACH, R. N. REUSCH, *Proc. Natl Acad. Sci. USA* **1997**, *94*, 9075–9079.

40 S. DAS, R. N. REUSCH, *Biochemistry* **2001**, *40*, 2075–2079.

41 D. SEEBACH, U. BRÄNDLI, P. SCHNURRENBERGER, M. PRZYBYLSKI, *Helv. Chim. Acta*, **1988**, *71*, 155–167.

42 D. SEEBACH, H. M. MÜLLER, H. M. BÜRGER, D. A. PLATTNER, *Angew. Chem. Int. Ed. Engl.* **1992**, *31*, 434–435.

43 D. A. PLATTNER, A. BRUNNER, M. DOBBLER, H.-M. MÜLLER, W. PETTER, P. ZBINDEN, D. SEEBACH, *Helv. Chim. Acta* **1993**, *76*, 2004–2033.

44 D. SEEBACH, H. M. BÜRGER, H.-M. MÜLLER, U. D. LENGWEILER, A. K. BECK, K. E. SYKES, P. A. BARKER, P. J. BARHAM, *Helv. Chim. Acta* **1994**, *77*, 1099–1123.

45 D. SEEBACH, T. HOFFMANN, F. N. M. KÜHNLE, U. D. LENGWEILER, *Helv. Chim. Acta* **1994**, *77*, 2007–2034.

46 P. WASER, M. RUEPING, D. SEEBACH, E. DUCHARDT, H. SCHWALBE, *Helv. Chim. Acta* **2001**, *84*, 1821–1845.

47 M. RUEPING, A. DIETRICH, V. BUSCHMANN, M. G. FRITZ, M. SAUER, D. SEEBACH, *Macromolecules* **2001**, *34*, 7042–7048.

48 M. ALBERT, D. SEEBACH, E. DUCHARDT, H. SCHWALBE, *Helv. Chim. Acta* **2002**, *85*, 633–658.

49 J. CORNIBERT, R. H. MARCHESSAULT, *Macromolecules* **1975**, *8*, 296–305.

50 S. BRÜCKNER, S. V. MEILLE, L. MALPEZZI, A. CESARO, L. NAVARINI, R. TOMBOLINI, *Macromolecules* **1988**, *21*, 967–972.

51 R. J. PAZUR, P. J. HOCKING, S. RAYMOND, R. H. MARCHESSAULT, *Macromolecules* **1998**, *31*, 6585–6592.

52 R. H. MARCHESSAULT, K. OKAMURA, C. aJ. SU, *Macromolecules* **1970**, *3*, 735–740.

53 J. CORNIBERT, R. H. MARCHESSAULT, H. BENOIT, G. WEIL, *Macromolecules* **1970**, *3*, 741–746.

54 S. H. GELLMAN, B. R. ADAMS, G. P. DADO, *J. Am. Chem. Soc.* **1990**, *112*, 460–461.

55 G. P. DADO, J. M. DESPER, S. H. GELLMAN, *J. Am. Chem. Soc.* **1990**, *112*, 8630–8632.

56 G.-B. LIANG, G. P. DADO, S. H. GELLMAN, *J. Am. Chem. Soc.* **1991**, *113*, 3994–3995.

57 S. H. GELLMAN, G. aP. DADO, G.-B. LIANG, B. R. ADAMS, *J. Am. Chem. Soc.* **1991**, *113*, 1164–1173.

58 G. P. DADO, J. M. DESPER, S. aK. HOLMGREN, C. J. RITO, S. H. GELLMAN, *J. Am. Chem. Soc.* **1992**, *114*, 4834–4843.

59 E. A. GALLO, S. H. GELLMAN, *J. Am. Chem. Soc.* **1993**, *115*, 9774–9788.

60 G. P. DADO, S. H. GELLMAN, *J. Am. Chem. Soc.* **1993**, *115*, 4228–4245.

61 G.-B. LIANG, J. M. DESPER, S. H. GELLMAN, *J. Am. Chem. Soc.* **1993**, *115*, 925–938.

62 E. A. GALLO, S. H. GELLMAN, *J. Am. Chem. Soc.* **1994**, *116*, 11560–11561.

63 G. P. DADO, S. H. GELLMAN, *J. Am. Chem. Soc.* **1994**, *116*, 1054–1062.

64 G. QUINKERT, E. EGERT, C. GRIESINGER, Macromolecular and supramolecular chemistry in *Aspects of Organic Chemistry, Structure*, G. QUINKERT, E. EGERT, C. GRIESINGER (Eds.), Verlag Helvetica Chimica Acta, Basel, **1996**.

65 M. NARITA, M. DOI, K. KUDO, Y. TERAUCHI, *Bull Chem Soc. Jpn.* **1986**, *59*, 3553–3557.

66 T. HINTERMAN, D. SEEBACH, *Synlett* **1997**, 437–438.

67 D. SEEBACH, J. L. MATTHEWS, *Chem Commun.* **1997**, 2015–2022.

68 G. CARDILLO, C. TOMASINI, *Chem Soc. Rev.* **1996**, 117–128.

69 T. C. BOGE, G. I. GEORG, The medicinal chemistry of β-amino acids: paclitaxel as an illustrative example in *Enantioselective Synthesis of β-amino acids*, E. JUARISTI (Ed.), Wiley-VCH, New York, **1997**.

70 For review, see: R. M. SCARBOROUGH, *Curr. Med. Chem.* **1999**, *6*, 971–981.

71 E. TAKASHIRO, I. HAYAKAWAA, T. NITTAA, A. KASUYAA, S. MIYAMOTOA, Y. OZAWAB, R. YAGIB, I. YAMAMOTOB, T. SHIBAYAMAC, A. NAKAGAWAC, Y. YABEA, *Bioorg. Med. Chem.* **1999**, *7*, 2063–2072 and references therein.

72 D. STEER, R. LEW, P. PERLMUTTER, A. I. SMITH, M. I. AGUILAR, *Biochemistry* **2002**, *41*, 10819–10826.

73 G. GUICHARD, A. ZERBIB, F. A. LE GAL, J. HOEBEKE, F. CONNAN, J. P. BRIAND, J. G. GUILLET, *J. Med. Chem.* **2000**, *43*, 3803–3808.

74 S. Reinelt, M. Marti, S. Dedier, T. Reitinger, G. Folkers, J. A. de Castro, D. Rognan, *J Biol. Chem.* **2001**, *276*, 24525–24530.

75 N. Koglin, C. Zorn, R. Beumer, C. Cabrele, C. Bubert, N. Sewald, O. Reiser, A. G. Beck-Sickinger, *Angew. Chem. Int. Ed.* **2003**, *42*, 202–205.

76 J. W. Trauger, E. E. Baird, P. B. Dervan, *Angew. Chem. Int. Ed.* **1998**, *37*, 1421–1423.

77 M. P. Glenn, M. J. Kelso, J. D. A. Tyndall, D. P. Fairlie, *J. Am. Chem. Soc.* **2003**, *125*, 640–641.

78 D. L. Steer R. A. Lew, P. Perlmutter, A. I. Smith, M.aI. Aguilar, *Curr. Med. Chem.* **2002** *9*, 811–822.

79 G. I. Georg, *Bioorg. Med. Chem. Lett.* **1993**, *3*, 2157.

80 J. R. Casimir, C. Didierjean, A. Aubry, M. Rodriguez, J.-P. Briand, G. Guichard, *Org. Lett.* **2000**, *2*, 895–897.

81 U. Diederichsen, H. W. Schmitt, *Angew. Chem. Int. Ed.* **1998**, *37*, 302–305.

82 A. M. Brückner, H. W. Schmitt, U. Diederichsen, *Helv. Chim. Acta* **2002**, *85*, 3855–3866.

83 G. Guichard, V. Semetey, C. Didierjean, A. Aubry, J.-P. Briand, M. Rodriguez, *J. Org. Chem.* **1999**, *64*, 8702–8705.

84 D. C. Cole, *Tetrahedron* **1994**, *50*, 9517–9582.

85 E. Juaristi, D. Quintana, J. Escalante, *Aldrichimica Acta* **1994**, *27*, 3–11.

86 E. Juaristi (Ed.), *Enantioselective Synthesis of β-Amino Acids*, Wiley-VCH, New York, **1997**.

87 N. Sewald, *Amino Acids* **1996**, *11*, 397–408.

88 E. Juaristi, H. Lopez-Ruiz. *Curr. Med. Chem.* **1999**, *6*, 983–1004.

89 S. Abele, D. Seebach, *Eur. J. Org. Chem.* **2000**, 1–15.

90 M. Liu, M. P. Sibi, *Tetrahedron* **2002**, *58*, 7991–8035.

91 J. A. Ellman, T. D. Owens, T. P. Tang, *Acc. Chem. Res.* **2002**, *35*, 984–995.

92 J. L. Matthews, M. Overhand, F. N. M. Kühnle, P. E. Ciceri, D. Seebach, *Liebigs Ann.* **1997**, 1371–1379.

93 K. Gademann, D. Seebach, *Helv. Chim. Acta* **2001**, *84*, 2924–2937.

94 K. Gademann, M. Ernst, D. Seebach, D. Hoyer, *Helv. Chim. Acta* **2000**, *83*, 16–33.

95 A. Jacobi, D. Seebach, *Helv. Chim. Acta* **1999**, *82*, 1150–1172.

96 J. L. Matthews, K. Gademann, B. Jaun, D. Seebach, *J. Chem. Soc., Perkin Trans 1* **1998**, 3331–3340.

97 S. Abele, G. Guichard, D. Seebach, *Helv. Chim. Acta* **1998**, *81*, 2141–2156.

98 G. Guichard, S. Abele, D. Seebach, *Helv. Chim. Acta* **1998**, *81*, 187–206.

99 K. Gademann, T. Kimmerlin, D. Hoyer, D. Seebach, *J. Med. Chem.* **2001**, *44*, 2460–2468.

100 A. Muller, C. Vogt, N. Sewald, *Synthesis* **1998**, 837–841.

101 A. Leggio, A. Liguori, A. Procopio, G. Sindona, *J. Chem. Soc., Perkin Trans 1* **1997**, 1969–1971.

102 H. Yang, K. Foster, C. R. J. Stephenson, W. Brown, E. Roberts, *Org. Lett.*, **2000**, *2*, 2177–2179.

103 D. Seebach, P. E. Ciceri, M. Overhand, B. Jaun, D. Rigo, L. Oberer, U. Hommel, R. Amstutz, H. Widmer, *Helv. Chim. Acta* **1996**, *79*, 2043–2066.

104 D. Seebach, S. Abele, K. Gademann, G. Guichard, T. Hintermann, B. Jaun, J. L. Matthews, J. V. Schreiber, L. Oberer, U. Hommel, H. Widmer, *Helv. Chim. Acta* **1998**, *79*, 932–982.

105 D. Seebach, J. V. Schreiber, S. Abele, X. Daura, W. F. van Gunsteren, *Helv. Chim. Acta* **2000**, *83*, 34–57.

106 D. Seebach, H. Estermann, *Tetrahedron Lett.* **1987**, *28*, 3103–3106.

107 J. Podlech, D. Seebach, *Liebigs Ann.* **1995**, 1217–1228.

108 D. Seebach, A. Jacobi, M. Rueping, K. Gademann, M. Ernst, B. Jaun, *Helv. Chim. Acta* **2000**, *83*, 2115–2140.

109 S. Krauthauser, L. A. Christianson, D. R. Powell, S. H. Gellman, *J. Am. Chem. Soc.* **1997**, *119*, 11719–11720.

110 I. A. Motorina, C. Huel, A. Quiniou, J. Mispelter, E. Adjadj, D. Grierson, *J. Am. Chem. Soc.* **2001**, *123*, 8–17.

111 K. Gademann, A. Häne, M. Rueping, B. Jaun, D. Seebach, *Angew. Chem. Int. Ed.* **2003**, *42*, 1534–1537.

112 R. A. Tromp, M. van der Hoeven, A. Amore, J. Brussee, M. Overhand, G. A. van der Marel, A. van der Gen, *Tetrahedron: Assymetry* **2001**, *12*, 1109–1112.

113 S. G. Davies, O. Ichihara, I. A. S. Walters, *J. Chem. Soc., Perkin Trans 1* **1994**, 1141–1147.

114 S. Hanessian, H. Yang, R. Schaum, *J. Am. Chem. Soc.* **1996**, *118*, 2507–2508.

115 S. Hanessian, H. Yang, *Tetrahedron Lett.*
1997, *38*, 3155–3158.

116 F.A. Davis, P. Zhou, B.-C. Chen, *Chem.
Soc. Rev.* **1998**, *27*, 13–18.

117 T.P. Tang, J.A. Ellman, *J. Org. Chem.*
1999, *64*, 12–13.

118 T.P. Tang, J.A. Ellman, *J. Org. Chem.*
2002, *67*, 7819–7832.

119 D.H. Appella, L.A. Christianson, D.A.
Klein, M.R. Richards, D.R. Powell,
S.H. Gellman, *J. Am. Chem. Soc.* **1999**,
121, 7574–7581.

120 P.R. LePlae, N. Umezawa, H.-S. Lee,
S.H. Gellman, *J. Org. Chem.* **2001**, *66*,
5629–5632.

121 X. Wang, J.F. Espinosa, S.H. Gellman,
J. Am. Chem. Soc. **2000**, *122*, 4821–4822.

122 H.-S. Lee, P.L. LePlae, E.A. Porter,
S.H. Gellman, *J. Org. Chem.* **2001**, *66*,
3597–3599.

123 H.-S. Lee, F.A. Syud, X. Wang, S.H.
Gellman, *J. Am. Chem. Soc.* **2001**, *123*,
7721–7722.

124 E.A. Porter, X. Wang, M.A. Schmitt,
S.H. Gellman, *Org. Lett.* **2002**, *4*, 3317–
3319.

125 M.G. Woll, J.D. Fisk, P.R. LePlae, S.H.
Gellman, *J. Am. Chem. Soc.* **2002**, 124,
12447–12452.

126 D.H. Appella, L.A. Christianson, I.L.
Karle, D.R. Powell, S.H. Gellman, *J.
Am. Chem. Soc.* **1999**, *121*, 6206–6212.

127 P.R. LePlae, T.L. Raguse, S.H. Gellman, *J. Org. Chem.* **2000**, *65*, 4766–4769.

128 M. Schinnerl, J.K. Murray, J.M. Langenhan, S.H. Gellman, *Eur. J. Org.
Chem.* **2003**, 721–726.

129 J.D. Winkler, E.L. Piatnitski, J. Mehl-
mann, J. Kasparec, P.H. Axelsen, *Angew. Chem. Int. Ed.* **2001**, *40*, 743–745.

130 T.A. Martinek, G.K. Tóth, E. Vass, M.
Hollósi, F. Fülöp, *Angew. Chem. Int. Ed.*
2002, *41*, 1718–1720.

131 F. Fülöp, *Chem. Rev.* **2001**, *101*, 2181–2204.

132 C. Cimarelli, G. Palmieri, *J. Org.
Chem.* **1996**, *61*, 5557–5563.

133 D. Xu, K. Prasad, O. Repiè, T.J. Black-
lock, *Tetrahedron: Asymmetry* **1997**, *8*,
1445–1451.

134 P. Michuh, D. Seebach, *Helv. Chim.
Acta* **2002**, *85*, 1567–1577.

135 D.A. Evans, F. Urpi, T.C Sommers, S.C
Clark, M.T Bilodeau, *J. Am. Chem. Soc.*
1990 *112*, 8215–8216.

136 D. Seebach, A. Boog, W.B. Schweizer,
Eur. J. Org. Chem. **1999**, 335–360.

137 V.M. Gutiérrez-García, G. Reyel-Ran-
gel, E. Juaristi, *Tetrahedron* **2001**, *57*,
6487–6496.

138 V.M. Gutiérrez-García, G. Reyel-Ran-
gel, O. Muoz-Muiz, E. Juaristi, *Helv.
Chim. Acta* **2002**, *85*, 4189–4199.

139 R. Ponsinet, G. Chassaing, J. Vaisser-
man, S. Lavielle, *Eur. J. Org. Chem.*
2000, 83–90.

140 G. Nagula, V.J. Huber, C. Lum, B.A.
Goodman, *Org. Lett.* **2000**, *2*, 3527–3529.

141 E. Arvanitis, H. Ernst, A.A. Ludwig
(née D'Souza), A.J. Robinson, P.B.
Wyatt, *J. Chem. Soc., Perkin Trans. 1*
1998, 521–528.

142 M.P. Sibi, P.K. Deshpande, *J. Chem.
Soc., Perkin Trans. 1* **2000**, 1461–1466.

143 D.A. Evans, L.D. Wu, J.J.M. Wiener,
J.S. Johnson, D.H.B. Ripin, J.S. Te-
drow, *J. Org. Chem.* **1999**, *64*, 6411–6417.

144 H.-S. Lee, J.-S. Park, B.M. Kim, S. Gellman, *J. Org. Chem.* **2003**, *68*, 1575–1578.

145 D. Seebach, S. Abele, T. Sifferlen, M.
Hänggi, S. Gruner, P. Seiler, *Helv.
Chim. Acta* **1998**, *81*, 2218–2219.

146 J.L. Matthews, C. Braun, C. Guibour-
denche, M. Overhand, D. Seebach,
Preparation of enantiopure β-amino acids
from α-amino acids using the Arndt–Eistert homologation, in E. Juaristi (Ed.),
Enantioselective Synthesis of β-Amino Acids,
Wiley-VCH, New York, 1997.

147 D. Seebach, T. Sifferlen, P.A.
Mathieu, A.M. Häne, C.M. Krell, D.J.
Bierbaum, S. Abele, *Helv. Chim. Acta*
2000, *83*, 2849–2864.

148 W. Traub, U. Shmueli, *Nature* **1963**,
198, 1165–1168.

149 P.M. Cowan, S. McGavin, *Nature* **1955**,
176, 501–503.

150 C.M. Deber, F.A. Bovey, J.P. Carver,
E.R. Blout, *J. Am. Chem. Soc.* **1970**, *92*,
6191–6198.

151 K. Kirshenbaum, A.E. Barron, R.A.
Goldsmith, P. Armand, E.K. Bradley,
K.T. Truong, K.A. Dill, F.E. Cohen,
R.N. Zuckermann, *Proc. Natl Acad. Sci.
USA* **1998**, *95*, 4303–4308.

152 P. Armand, K. Kirshenbaum, R.A.
Goldsmith, S. Farr-Jones, A.E. Bar-
ron, K.T. Truong, K.A. Dill, D.F.
Mierke, F.E. Cohen, R.N. Zucker-

MANN, E. K. BRADLEY, *Proc. Natl Acad. Sci. USA* **1998**, *95*, 4309–4314.

153 S. ABELE, K. VÖGTLI, D. SEEBACH, *Helv. Chim. Acta* **1999**, *82*, 1539–1558.

154 Y. J. CHUNG, B. R. HUCK, L. A. CHRISTIANSON, H. E. STANGER, S. KRAUTHAUSER, D. R. POWELL, S. H. GELLMAN, *J. Am. Chem. Soc.* **2000**, *122*, 3995–4004.

155 B. R. HUCK, J. M. LANGENHAN, S. H. GELLMAN, *Org Lett.* **1999**, *1*, 1717–1720.

156 A. M. ACKERMAN, D. K. DE JONGH, H. VELDSTRA, *Recl. Trav. Chim. Pays-Bas* **1951**, *70*, 899–916.

157 S. I. KLEIN, M. CZEKAJ, B. F. MOLINO, V. CHU, *Bioorg. Med. Chem. Lett.* **1997**, *7*, 1773–1778.

158 A. BANERJEE, P. BALARAM, *Curr. Sci.* **1997**, *73*, 1067–1077.

159 J. J. BARCHI JR, X. HUANG, D. H. APELLA, L. A. CHRISTIANSON, S. R. DURELL, S. GELLMAN, *J. Am. Chem. Soc.* **2000**, *122*, 2711–2718.

160 D. H. APPELLA, L. A. CHRISTIANSON, D. A. KLEIN, D. R. POWELL, X. HUANG, J. J. BARCHI, S. H. GELLMAN, *Nature* **1997**, *387*, 381–384.

161 D. SEEBACH, K. GADEMANN, J. V. SCHREIBER, J. L. MATTHEWS, T. HINTERMANN, B. JAUN, L. OBERER, U. HOMMEL, H. WIDMER, *Helv. Chim. Acta* **1997**, *80*, 2033–2038.

162 M. RUEPING, J. V. SCHREIBER, G. LELAIS, B. JAUN, D. SEEBACH, *Helv. Chim. Acta* **2002**, *85*, 2577–2593.

163 S. ABELE, P. SEILER, D. SEEBACH, *Helv. Chim. Acta* **1999**, *82*, 1559–1571.

164 K. GADEMANN, B. JAUN, D. SEEBACH, R. PEROZZO, L. SCAPOZZA, G. FOLKERS, *Helv. Chim. Acta* **1999**, *82*, 1–11.

165 D. H. APPELLA, J. J. BARCHI JR, S. R. DURELL, S. H. GELLMAN, *J. Am. Chem. Soc.* **1999**, *121*, 2309–2310.

166 X. DAURA, K. GADEMANN, B. JAUN, D. SEEBACH, W. F. VAN GUNSTEREN, A. E. MARK, *Angew. Chem. Int. Ed. Engl.* **1999**, *38*, 236–240.

167 R. FAIRMAN, K. R. SHOEMAKER, E. J. YORK, J. STEWART, R. L. BALDWIN, *Proteins: Struct. Funct. Genet.* **1989**, *5*, 1–7.

168 T. L. RAGUSE, J. R. LAI, S. H. GELLMAN, *Helv. Chim. Acta* **2002**, *85*, 4154–4164.

169 Y.-D. WU, D.-P. WANG, *J. Am. Chem. Soc.* **1998**, *120*, 13485–13493.

170 D. SEEBACH, Y. R. MAHAJAN, R. SENTHILKUMAR, M. RUEPING, B. JAUN, *Chem. Commun.* **2003**, 1598–1599.

171 M. RUEPING, B. JAUN, D. SEEBACH, *Chem. Commun.* **2000**, 2267–2268.

172 A. GLÄTTLI, X. DAURA, D. SEEBACH, W. F. VAN GUNSTEREN, *J. Am. Chem. Soc.* **2002**, *124*, 12972–12978.

173 B. W. GUNG, D. ZOU, A. M. STALCUP, C. E. COTTRELL, *J. Org. Chem.* **1999**, *64*, 2176–2177.

174 T. ETEZADY-ESFARJANI, C. HILTY, K. WÜTHRICH, M. RUEPING, J. SCHREIBER, D. SEEBACH, *Helv. Chim. Acta* **2002**, *85*, 1197–1209.

175 T. L. RAGUSE, E. A. PORTER, B. WEISBLUM, S. H. GELLMAN, *J. Am. Chem. Soc.* **2002**, *124*, 12774–12785.

175a P. I. ARVIDSSON, M. RUEPING, D. SEEBACH, *Chem. Commun.* **2001**, 649–650.

175b R. P. CHENG, W. F. DEGRADO, *J. Am. Chem. Soc.* **2001**, *123*, 5162–5163.

176 T. L. RAGUSE, J. R. LAI, S. H. GELLMAN, *J. Am. Chem. Soc.* **2003**, *125*, 5592–5593.

177 S. A. HART, A. B. F. BAHADOOR, E. E. MATTHEWS, X. J. QIU, A. SCHEPARTZ, *J. Am. Chem. Soc.* **2003**, *125*, 4022–4023.

178 T. L. RAGUSE, J. R. LAI, P. R. LEPLAE, S. GELLMAN, *J. Am. Chem. Soc.* **2001**, *3*, 3963–3966.

179 R. P. CHENG, W. F. DEGRADO, *J. Am. Chem. Soc.* **2002**, *124*, 11564–11565.

180 Y. D. WU, D. P. WANG, *J. Am. Chem. Soc.* **1998**, *121*, 9352–9362.

181 K. MÖHLE, R. GUNTHER, M. THORMANN, N. SEWALD, H. J. HOFMANN, *Biopolymers* **1999**, *50*, 167–183.

182 R. BARON, D. BAKOWIES, W. E. VAN GUNSTEREN, X. DAURA, *Helv. Chim. Acta* **2002**, *85*, 3872–3882.

183 CHRISTIANSON, *J. Comput. Chem.* **2000**, *21*, 763–773.

184 J. APPLEQUIST, K. A. BODE, D. H. APPELLA, L. A. CHRISTIANSON, S. H. GELLMAN, *J. Am. Chem. Soc.* **1998**, *120*, 4891–4892.

184a P. R. LEPLAE, J. D. FISK, E. A. PORTER, B. WEISBLUM, S. H. GELLMAN, *J. Am. Chem. Soc.* **2002**, *124*, 6820–6821.

184b J.-S. PARK, H.-S. LEE, J. R. LAI, B. M. KIM, S. H. GELLMAN, *J. Am. Chem. Soc.* **2003**, *125*, 8539–8545.

185 F. GUERITTE-VOEGELEIN, D. GUENARD, L. MANGATAL, P. POTIER, J. GUILHEM, M. CESARIO, C. PASCAL, *Acta Crystallogr., Sect. C*, **1990**, *46*, 781–784.

186 Q. GAO, S. H. CHEN, *Tetrahedron Lett.* **1996**, *37*, 3425–3428.

187 M. Doi, Y. In, M. Inoue, T. Ishida, K. Lisuka, K. Akahane, H. Harada, H. Umeyama, Y. Kiso, *J. Chem. Soc. Perkin Trans. 1* **1991**, 1153–1159.

188 J.R. Peterson, H.D. Do, R.D. Rogers, *Pharmaceut. Res.* **1991**, 8, 908–912.

189 R.W. Miller, R.G. Powell, C.R. Smith Jr, E. Arnold, J. Clardy, *J. Org. Chem.* **1981**, 46, 1469–1474.

190 S.A. Monti, *J. Org. Chem.* **1970**, 35, 380–383.

191 D. Seebach, S. Abele, Karl Gademann, B. Jaun, *Angew. Chem.* **1999**, 111, 1700–1703; *Angew. Chem. Int.* **1999**, 38, 1595–1597.

192 I. Karle, H.N. Gopi, P. Balaram, *Proc. Natl Acad. Sci. USA* **2002**, 99, 5160–5164.

193 Y.J. Chung, L.A. Christianson, H.E. Stanger, D.R. Powell, S.H. Gellman, *J. Am. Chem. Soc.* **1998**, 120, 10555–10556.

194 X. Daura, K. Gademann, H. Schäfer, B. Jaun, D. Seebach, W.F. van Gunsteren, *J. Am. Chem. Soc.* **2001**, 123, 2393–2404.

195 S. Abele, Dissertation, ETH-Zürich, No. 13203, 1999, p. 94.

196 J.M. Langenhan, I. A. Guzei, S.H. Gellman, *Angew. Chem. Int. Ed.* **2003**, 42, 2402–2405.

197 I. Karle, H.N. Gopi, P. Balaram, *Proc. Natl Acad. Sci. USA* **2001**, 98, 3716–3719.

198 H.N. Gopi, R.S. Roy, S.R. Raghothama, I. Karle, P. Balaram, *Helv. Chim. Acta* **2002**, 85, 3313–3330.

199 B.R. Huck, J.D. Fisk, S.H. Gellman, *Org. Lett.* **2000**, 2, 2607–2610.

200 T. Hintermann, K. Gademann, B. Jaun, D. Seebach, *Helv. Chim. Acta* **1998**, 81, 983–1002.

201 S. Hanessian, X. Luo, R. Schaum, S. Michnick, *J. Am. Chem. Soc.* **1998**, 120, 8569–8570.

202 S. Hanessian, X. Luo, R. Schaum, *Tetrahedron Lett.* **1999**, 40, 4925–4929.

203 M. Brenner, D. Seebach, *Helv. Chim. Acta* **2001**, 84, 2155–2166.

204 D. Seebach, L. Schaeffer, M. Brenner, D. Hoyer, *Angew. Chem. Int. Ed.* **2003**, 42, 776–778.

205 M.G. Woll, J.R. Lai, I.A. Guzei, S.J.C. Taylor, M.E.B. Smith, S.H. Gellman, *J. Am. Chem. Soc.* **2001**, 123, 11077–11078.

206 D. Seebach, M. Brenner, M. Rueping, B. Schweizer, B. Jaun, *Chem. Commun.* **2001**, 207–208.

207 D. Seebach, M. Brenner, M. Rueping, B. Jaun, *Chem. Eur. J.* **2002**, 8, 573–584.

208 M. Hagihara, N.J. Anthony, T.J. Stout, J. Clardy, S.L. Schreiber, *J. Am. Chem. Soc.* **1992**, 114, 6568–6570.

209 P. Coutrot, C. Grison, S. Genève, C. Didierjean, A. Aubry, A. Vicherat, M. Marraud, *Lett. Pept. Sci.* **1997**, 4, 415–422.

210 M. Brenner, D. Seebach, *Helv. Chim. Acta* **2001**, 84, 1181–1189.

211 H.N. Rydon, *J. Chem. Soc.* **1964**, 1328–1333.

212 D. Zanuy, C. Alemán, S. Muñoz-Guerra, *Int. J. Biol. Macromol.* **1998**, 23, 175–184.

213 J. Puiggal, S. Muñoz-Guerra, A. Rodríguez-Galán, C. Alegre, J.A. Subirana, *Makromol. Chem. Macromol. Symp.* **1988**, 20/21, 167.

214 J. Melis, D. Zanuy, C. Alemán, M. García-Alvarez, S. Muñoz-Guerra, *Macromolecules* **2002**, 35, 8774–8780.

215 R.W. Hoffmann, *Angew. Chem.* **1992**, 104, 1147–1157; *Angew. Chem. Int. Ed. Engl.* **1992**, 31, 1124–1134.

216 R.W. Hoffman, *Angew. Chem. Int. Ed.* **2000**, 39, 2054–2070.

217 R.W. Hoffman, M.A. Lazaro, F. Caturla, E. Framery, I. Valancogne, C.A.G.N. Montalbetti, *Tetrahedron Lett.* **2000**, 40, 5983–5986.

218 R.W. Hoffman, F. Caturla, M.A. Lazaro, E. Framery, M.C. Bernabeu, I. Valancogne, C.A.G.N. Montalbetti, *New. J. Chem.* **2000**, 24, 187–194.

219 D.L. Boger, T.M. Ramsey, H. Cai, S.T. Hoehn, J. Stubbe, *J. Am. Chem. Soc.* **1998**, 120, 9149–9158.

220 Y. Kato, N. Fusetani, S. Matsunaga, K. Hashimoto, K. Koseki, *J. Org. Chem.* **1988**, 53, 3930–3932.

221 C. Baldauf, R. Günther, H.-J. Hofmann, *Helv. Chim. Acta* **2003**, 86, 2573–2588.

222 H. Kogen, T. Nishi, *J. Chem. Soc. Chem. Commun.* **1987**, 311–312.

223 M.T. Reetz, D. Röhrig, *Angew. Chem.* **1989**, 101, 1732–1734; *Angew. Chem. Int. Ed. Engl.* **1989**, 28, 1706–1709.

224 S. Hanessian, R. Schaum, *Tetrahedron Lett.* **1997**, 38, 163–166.

225 M. Smrcina, P. Majer, E. Majerova, T.A. Guerassina, M.A. Eisenstat, *Tetrahedron* **1997**, 53, 12867–12874.

226 P. Buchschacher, J.-M. Cassal, A. Fürst, W. Meier, *Helv. Chim. Acta,* **1977**, 60, 2747–2755.

227 C.C. TSENG, S. TERASHIMA, S. YAMADA, *Chem. Pharm. Bull.* **1977**, *25*, 29–40.

228 A.M.P. KOSKINEN, P.M. PIHKO, *Tetrahedron Lett.* **1994**, *35*, 7417–7420.

229 P.M. PIHKO, A.M.P. KOSKINEN, *J. Org. Chem.* **1998**, *63*, 92–98.

230 M. BRENNER, D. SEEBACH, *Helv. Chim. Acta* **1999**, *82*, 2365–2379.

231 N.H. ANDERSEN, J.W. NEIDIGH, S.M. HARRIS, G.M. LEE, Z. LIU, H. TONG, *J. Am. Chem. Soc.* **1997**, *119*, 8547–8561.

232 T. HINTERMANN, D. SEEBACH, *Chimia* **1997**, *51*, 244–247.

233 J. FRACKENPOHL, P.I. ARVIDSSON, J.V. SCHREIBER, D. SEEBACH, *Chembiochem* **2001**, *2*, 445–455.

234 E.A. PORTER, B. WEISBLUM, S.H. GELLMAN, *J. Am. Chem. Soc.* **2002**, *124*, 7324–7330.

235 H. WIEGAND, B. WIRZ, A. SCHWEITZER, G.P. CAMENISCH, M.I. RODRIGUEZ PEREZ, G. GROSS, R. WOESSNER, R. VOGES, P.I. ARVIDSSON, J. FRACKENPOHL, D. SEEBACH, *Biopharm. Drug Dispos.* **2002**, *23*, 251–262.

236 E. HABERMANN, *Science* **1972**, *177*, 314–322.

237 M. ZASLOFF, *Proc. Natl Acad. Sci. USA* **1987**, *84*, 5449–5453.

238 H. STEINER, D. HULTMARK, A. ENGSTROM, H. BENNICH, H.G. BOMAN, *Nature* **1981**, *292*, 246–248.

239 M. ZASLOFF, *Nature* **2002**, *415*, 389–395.

240 A. TOSSI, L. SANDRI, A. GIANGASPERO, *Biopolymers* **2000**, *55*, 4–30.

241 Z. OREN, Y. SHAI, *Biopolymers* **1998**, *47*, 451–463.

242 Y. SHAI, *Biopolymers* **2002**, *66*, 236–248.

243 M. DATHE, T. WIEPRECHT, *Biochim. Biophys. Acta* **1999**, *1462*, 71–87.

244 Y. HAMURO, J.P. SCHNEIDER, F. DeGRADO, *J. Am. Chem. Soc.* **1999**, *121*, 12200–12201.

245 D. LIU, W.F. DeGRADO, *J. Am. Chem. Soc.* **2001**, *123*, 7553–7559.

246 P.I. ARVIDSSON, J. FRACKENPOHL, N.S. RYDER, B. LIECHTY, F. PETERSEN, H. ZIMMERMAN, G.P. CAMENISCH, R. WOESSNER, D. SEEBACH, *ChemBiochem* **2001**, *2*, 771–773.

247 H.C. CHEN, J.H. BROWN, J.L. MORELL, C.M. HUANG, *FEBS Lett.* **1988**, *236*, 462–466.

248 E.A. PORTER, X. WANG, H.-S. LEE, B. WEISBLUM, S.H. GELLMAN, *Nature* **2000**, *404*, 565. (Erratum: *Nature* **2000**, *405*, 298).

249 A.T. ULIJASZ, A. GRENADER, B. WEISBLUM, *J. Bacteriol.* **1996**, *178*, 6305–6309.

250 M. WERDER, H. HAUSER, S. ABELE, D. SEEBACH, *Helv. Chim. Acta* **1999**, *82*, 1774–1783.

251 P. BRAZEAU, W. VALE, R. BURGUS, N. LING, M. BUTCHER, J. RIVIER, R. GUILLEMIN, *Science* **1973**, *179*, 77–79.

252 D. HOYER, G.I. BELL, M. BERELOWITZ, J. EPELBAUM, W. FENIUK, P.P.A. HUMPHREY, A-M. O'CARROLL, Y.C. PATEL, A. SCHONBRUNN, J.E. TAYLOR, T. REISINE, *Trends Pharmacol. Sci.* **1995**, *16*, 86–88.

253 W. BAUER, U. BRINER, W. DOEPFNER, R. HALLER, R. HUGUENIN, P. MARBACH, T.J. PETCHER, J. PLESS, *Life Sci.* **1982**, *13*, 1133–1140.s

254 D. SEEBACH, M. RUEPING, P.I. ARVIDSSON, T. KIMMERLIN, P. MICUCH, C. NOTI, D. LANGENEGGER, D. HOYER D, *Helv. Chim. Acta* **2001**, *84*, 3503–3510.

255 C. NUNN, M. RUEPING, D. LANGENEGGER, E. SCHUEPBACH, T. KIMMERLIN, P. MICUCH, K. HURTH, D. SEEBACH, D. HOYER, *Naunyn-Schmiedeberg's Arch. Pharmacol.* **2003**, *367*, 578–587.

256 A. PROCHIANTZ, *Curr. Opin. Cell Biol.* **2000**, *12*, 400–406.

257 S.R. SCHWARZE, K.A. HRUSKA, S,F. DOWDY, *Trends Cell. Biol.* **2000**, *10*, 290.

258 M. LINDGREN, M. HALLBRINK, A. PROCHIANTZ, U. LANGEL, *Trends Pharmacol. Sci.* **2000**, *21*, 99–103.

259 P.A. WENDER, D.J. MITCHELL, K. PATTABIRAMAN, E.T. PELKEY, L. STEINMAN, J.B. ROTHBARD, *Proc. Natl Acad. Sci. USA* **2000**, *97*, 13003–13008.

260 N. UMEZAWA, M.A. GELMAN, M.C. HAIGIS, R.T. RAINES, S.H. GELLMAN, *J. Am. Chem. Soc.* **2002**, *124*, 368–369.

261 M. RUEPING, Y. MAHAJAN, M. SAUER, D. SEEBACH, *ChemBiochem* **2002**, *2–3*, 257–259.

262 C. GARCÍA-ECHEVERRÍA, S. RUETZ, *Bioorg. Med. Chem. Lett.* **2003**, *13*, 247–251.

263 T. SIFFERLEN, M. RUEPING, K. GADEMANN, B. JAUN, D. SEEBACH, *Helv. Chim. Acta* **1999**, *82*, 2067–2093.

264 M. GUDE, U. PIARULLI, D. POTENZA, B. SALOM, C. GENNARI, *Tetrahedron Lett.* **1996**, *37*, 8589–8592.

265 C. GENNARI, M. GUDE, D. POTENZA, U. PIARULLI, *Chem. Eur. J.* **1998**, *4*, 1924–1931.

266 W.J. MOREE, G. VAN DER MAREL, R.J. LISKAMP, *J. Org. Chem.* **1995**, *60*, 5157–5169.

267 R. GÜNTHER, H.-J. HOFFMANN, *J. Am. Chem. Soc.* **2001**, *123*, 247–255.

268 G. LELAIS, D. SEEBACH, Proceedings of the 2nd International and the 17th American Peptide Symposium – San Diego (June 2001), in *Peptides: The Wave of the Future*, M. LEBL, R.A. HOUGHTEN (Eds), American Peptide Society, San Diego, **2001**, pp. 581–582.

269 D. YANG, F.-F. NG, Z.-J. LI, Y.-D. WU, K.W.-K. CHAN, D.-P. WANG, *J. Am. Chem. Soc.* **1996**, *118*, 9794–9795.

270 D. YANG, J. QU, B. LI, F.-F. NG, X.-C. WANG, K.-K. CHEUNG, D.-P. WANG, Y.-D. WU, *J. Am. Chem. Soc.* **1999**, *121*, 589–590.

271 K. BURGESS, D.S. LINTHICUM, H. SHIN, *Angew. Chem. Int. Ed. Engl.* **1995**, *34*, 907–908.

272 K. BURGESS, J. IBARZO, D.S. LINTHICUM, D.H. RUSSELL, H. SHIN, A. SHITANG-KOON, R. TOTANI, A.J. ZHANG, *J. Am. Chem. Soc.* **1997**, *119*, 1556–1564.

273 V. SEMETEY, D. ROGNAN, C. HEMMERLIN, R. GRAFF, J.-P. BRIAND, M. MARRAUD, G. GUICHARD, *Angew. Chem.* **2002**, *115*, 1973–1975; *Angew. Chem. Int. Ed.* **2002**, *41*, 1893–1895.

274 C. HEMMERLIN, M. MARRAUD, D. ROGNAN, R. GRAFF, V. SEMETEY, J.P. BRIAND, G. GUICHARD, *Helv. Chim. Acta* **2002**, *85*, 3692–3711.

275 C.Y. CHO, E.J. MORAN, S.R. CHERRY, J.C. STEPHANS, S.P.A. FODOR, C.L. ADAMS, A. SUNDARAM, J.W. JACOBS, P.G. SCHULTZ, *Science* **1993**, *261*, 1303–1305.

276 C.Y. CHO, R.S. YOUNGQUIST, S.J. PAIKOFF, M.H. BERESINI, A.R. HERBERT, L.T. BERLEAU, C.W. LIU, D.E. WEMMER, T. KEOUGH, P.G. SCHULTZ, *J. Am. Chem. Soc.* **1998**, *120*, 7706.

277 C. GENNARI, B. SALOM, D. POTENZA, A. WILLIAMS, *Angew. Chem. Int. Ed. Engl.* **1994**, *33*, 2067–2069.

278 C. GENNARI, C. LONGARI, S. RESSEL, B. SALOM, U. PIARULLI, S. CACCARELLI, A. MIELGO, *Eur. J. Org. Chem.* **1998**, 2437–2449.

279 D. YANG, Y.-H. ZHANG, N.Y. ZHU, *J. Am. Chem. Soc.* **2002**, *124*, 9966–9967.

280 E. TESTA, B.J.R. NICOLAUS, L. MARIANI, G. PAGANI, *Helv. Chim. Acta* **1963**, *46*, 766–780.

281 D. YANG, B. LI, F.-F. NG, Y.-L. YAN, J. QU, Y.-D. WU, *J. Org. Chem.* **2001**, *66*, 7303–7312.

282 A. MATSUNGA, K. MURAAKAMI, H. NOHIRA, M. OHASHI, I. YAMAMOTO, *Chirality* **1993**, *5*, 41–48.

283 I. SHIN, M.-R. LEE, J. LEE, M. JUNG, W. LEE, J. YOON, *J. Org. Chem.* **2000**, *65*, 7667–7675.

284 Y.-D. WU, D.-P. WANG, K.W.K. CHAN, D. YANG, *J. Am. Chem. Soc.* **1999**, *121*, 11189–11196.

285 C. PETER, X. DAURA, W.F. VAN GUNSTEREN, *J. Am. Chem. Soc.* **2000**, *122*, 7461–7466.

286 J.M. KIM, Y. BI, S. PAIKOFF, P.G. SCHULTZ, *Tetrahedron Lett.* **1996**, *37*, 5305–5308.

287 A. BOEIJEN, R.M.J. LISKAMP, *Eur. J. Org. Chem.* **1999**, 2127–2135.

288 G. GUICHARD, V. SEMETEY, M. RODRIGUEZ, J.P. BRIAND, *Tetrahedron Lett.* **2000**, *41*, 1553–1557.

289 H. RINK, *Tetrahedron Lett.* **1987**, *28*, 3787–3790.

290 D. SEEBACH, L. SCHAEFFER, F. GESSIER, P. BINDSCHÄDLER, C. JÄGER, D. JOSIEN, S. KOPP, G. LELAIS, Y.R. MAHAJAN, P. MICUCH, R. SEBESTA, B.W. SCHWEIZER, Helv. Chim. Acta **2003**, *86*, 1852–1861.

291 T.A. MARTINEK, F. FÜLÖP, *Eur. J. Biochem.* **2003**, *270*, 3657–3666.

292 F. ROSSI, G. LELAIS, D. SEEBACH, Helv. Chim. Acta **2003**, *86*, 2653–2661.

293 T. KIMMERLIN, K. NAMOTO, D. SEEBACH, Helv. Chim. Acta **2003**, *86*, 2104–2109.

294 A.M. BRÜCKNER, P. CHAKRABORTY, S.H. GELLMAN, U. DIEDERICHSEN, Angew. Chem. Int. Ed. **2003**, *42*, 4395–4399.

295 M.A. GELMAN, S. RICHTER, H. CAO, N. UMEZAWA, S.H. GELLMAN, T.M. RANA, Org. Lett. **2003**, *5*, 3563–3565.

296 T.B. POTOCKY, A.K. MENON, S.H. GELLMAN, J. Biol. Chem. **2003**, in press.

3
Regulation of Gene Expression
with Pyrrole-Imidazole Polyamides

Peter B. Dervan, Eric J. Fechter, Benjamin S. Edelson, and Joel M. Gottesfeldr

3.1
Introduction

The natural product distamycin contains three *N*-methylpyrrole amino acids and binds in the minor groove of DNA at A,T tracts 4–5 base pairs (bp) in size [1, 2]. Distamycin inhibits DNA-dependent processes, including transcription, and has antibacterial [3], antimalarial [4], antifungal [5] and antiviral activities [6], but is of limited use because of toxicity [7]. Efforts to bring distamycin analogs to the clinic have focused on anti-infective therapeutics [8–11]. These compounds have been optimized for pathogen activity and pharmacological properties, with DNA binding specificity not providing a major driving force in ligand selection.

Dickerson, Rich and Wemmer revealed by X-ray and NMR structural studies that the crescent-shaped molecule could bind A,T tracts in both 1:1 and 2:1 li-

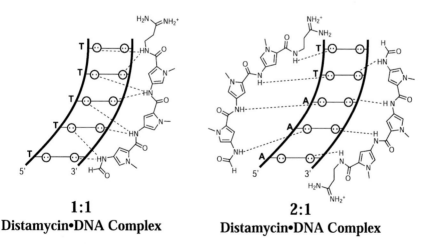

1:1
Distamycin•DNA Complex

2:1
Distamycin•DNA Complex

Fig. 3.1 Schematic representation of the two modes of distamycin:DNA complexes with putative hydrogen bonds shown as dashed lines. Circles with dots represent lone pairs of N(3) of purines and O(2) of pyrimidines

Pseudo-peptides in Drug Discovery. Edited by Peter E. Nielsen
Copyright © 2004 Wiley-VCH Verlag GmbH & Co. KGaA, Weinheim
ISBN: 3-527-30633-1

gand:DNA stoichiometries (Fig. 3.1) [12–14]. Informed by these structures, we explored whether designed distamycin analogs could be tuned by chemical modification in a predictable fashion to bind a very large number of different DNA sequences, i.e. create an artificial small molecule language to read the minor groove digitally, similar in function to nature's proteins [15]. This might underpin a rational chemical approach to the regulation of gene expression by chemical methods.

After a 20-year search, we demonstrated that synthetic analogs of the N-methyl-pyrrole (Py) carboxamide ring afford a set of heterocycles that can be combined – as *unsymmetrical ring pairs* – in a modular fashion to recognize specifically a large repertoire of DNA sequences with affinities and specificities comparable to DNA-binding proteins [16]. In this chapter we describe advances in the field of DNA-binding polyamides, cellular and nuclear uptake properties and recent biological applications.

3.2
Pairing Rules

In a formal sense, the four Watson–Crick base pairs can be differentiated on the minor groove floor by the specific positions of hydrogen bond donors and acceptors, as well as by differences in molecular shape [16]. The exocyclic NH_2 of guanine presents a bump on the edge of a G,C base pair, whereas a T,A base pair presents a cleft. A key study in the early 1990s demonstrated that the N-methylimidazole-(Im)-containing polyamide ImPyPy bound to the 5-bp sequence 5′-WGWCW-3′ (where W=A or T) [17]. This result was rationalized in terms of the formation of a 2:1 polyamide–DNA complex, subsequently verified by NMR [18], in which an antiparallel ring pairing of Im stacked against Py could specifically distinguish a G,C from a C,G base pair. This discovery pointed toward a new paradigm of unsymmetrical ring pairs for specific recognition in the DNA minor groove (Figs. 3.2 and 3.3).

The Im/Py pair has been explored by extensive studies, including analyses of binding in hundreds of different sequence contexts. Crystal structures confirmed the existence of a hydrogen bond between the Im nitrogen and the exocyclic NH_2 of guanine when the Im/Py pair binds opposite the G,C base pair [19]. The preference for a linear hydrogen bond, coupled with the unfavorable angle to an Im over the cytosine side of the base pair, provides a basis for the ability of an Im/Py pair to specifically discriminate G,C from C,G. These crystal structures also revealed other key ligand–DNA interactions, such as a series of hydrogen bonds between the amide groups of the polyamides and the edges of the bases on the adjacent DNA strand. Thermodynamic investigations dissected binding free energies into enthalpic and entropic contributions, revealing that the sequence selectivity of the Im/Py pair is driven by a favorable enthalpic contribution [20].

Within the context of Watson–Crick base pair recognition by unsymmetrical heterocyclic ring pairs, and informed by high resolution crystallographic data from a

Py/Im targets **C•G** ⟹

Py/Hp targets **A•T** ⟹

Hp/Py targets **T•A** ⟹

Im/Py targets **G•C** ⟹

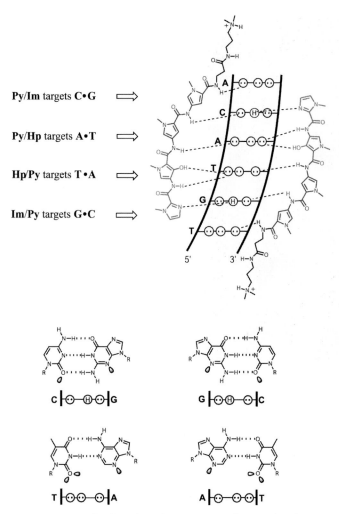

Fig. 3.2 Pairing rules for polyamide recognition of all four Watson–Crick base pairs of DNA. Putative hydrogen bonds are shown as dashed lines. Circles with dots represent lone pairs of N(3) of purines and O(2) of pyrimidines, and circles containing an H represent the 2-amino group of guanine. The R group represents the sugar-phosphate backbone of DNA, and shaded orbitals represent electron lone pairs projecting into the minor groove

polyamide–DNA complex, the N-methyl-3-hydroxypyrrole (Hp) monomer was designed as a thymine-selective recognition element when paired across from Py (Figs. 3.2 and 3.3) [21]. It was anticipated for steric reasons that a substituent such as hydroxyl would not "fit" opposite A, but would be accommodated at T. Crystal structures of two different Hp-containing polyamides, as their 2:1 complexes with DNA, have been determined at high resolution [22, 23]. An Hp/Py pair was shown to distinguish T,A from A,T, G,C, and C,G base pairs using a combination

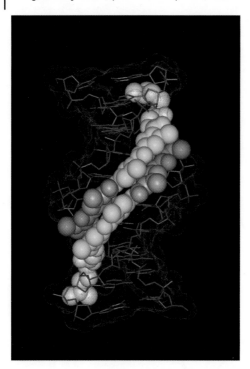

Fig. 3.3 X-ray crystal structure of Im-HpPyPy-β-Dp (Dp=dimethylamino-propylamine) bound in a 2:1 complex with its target DNA site, 5′-AGTACT-3′

of specific hydrogen bonds between the hydroxyl and the thymine-O2, along with shape selective recognition of an asymmetric cleft between the thymine-O2 and adenine-C2 (Figs. 3.2 and 3.3). Hp polyamides bind with lower affinity than their Py counterparts [21]. Partial melting of the T,A base pair recognized by an Hp/Py ring pair was thought to be a possible explanation [22], but the distortion was observed in only one of two crystal structures. A consistent lengthening of the amide–DNA hydrogen bond on the C-terminal side of the Hp residue is observed in both structures, although this may be inadequate to account for the energetic differences [23]. A recent computational study argues that desolvation of the hydroxyl group upon insertion into the minor groove accounts for the energetic penalty [24]. Together, three rings – Py, Im, and Hp – can be combined as unsymmetrical pairs to recognize specifically each of the four Watson–Crick base pairs; Im/Py is specific for G,C and Hp/Py for T,A. These interactions can be conveniently described as pairing rules (Fig. 3.2). The pairing rules should be considered as guidelines only. Antiparallel polyamide dimers bind B-form DNA, and there are limitations regarding sequences targeted due to the sequence dependent microstructure of DNA.

dihedral required for hydrogen bonding. Additionally, close contacts of β-alanine to the floor of the minor groove provided a structural explanation for its observed A/T specificity.

3.6
Solid Phase Methods

The investigation of minor groove-binding polyamides was greatly accelerated by the implementation of solid-phase synthesis [48]. Originally demonstrated on Boc-β-Ala-PAM resin with Boc-protected monomers, it was also shown that Fmoc chemistry could be employed with suitably protected monomers and Fmoc-β-Ala-Wang resin (Fig. 3.8) [49]. Recently, Pessi and coworkers used a sulfonamide-based safety-catch resin to prepare derivatives of hairpin polyamides [50]. Upon activation of the linker, resin-bound polyamides were readily cleaved with stoichiometric quantities of nucleophile to provide thioesters or peptide conjugates.

While allowing rapid preparation of a range of polyamides, these resins install a T,A-selective β-Ala residue at the *C*-terminus, which places limits on the DNA sites that can be targeted [46]. The shortest tail available from these resins is a propanolamide, obtained by reductive cleavage. Polyamides prepared on Boc-Gly-PAM resin can be reductively cleaved to obtain ethanolamide tails, but it was expected that further truncation of the *C*-terminus would be necessary for tolerance

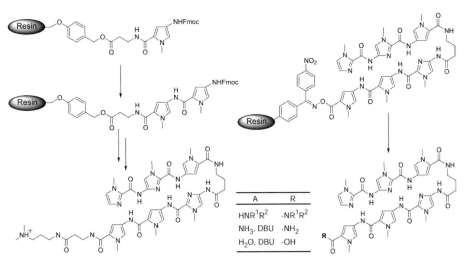

Fig. 3.8 Variations to solid phase synthesis of polyamides. Use of Fmoc monomers on β-Ala-Wang resin (left) provides polyamides containing a β-alanine residue near the *C*-termini. Polyamides synthesized on the Kaiser oxime resin (right) can have shorter *C*-terminal groups than molecules prepared on β-Ala-PAM resin. The amine HNR^1R^2 may be a primary or secondary alkyl amine

A	R
HNR^1R^2	$-NR^1R^2$
NH_3, DBU	$-NH_2$
H_2O, DBU	$-OH$

of G,C at the tail position [51]. The Kaiser oxime resin was therefore adapted to polyamide synthesis, allowing the preparation of polyamides with shorter C-termini (Fig. 3.8). These molecules display the desired tolerance for G,C bases while maintaining high affinities [52].

3.7
Benzimidizole Ring Systems for Minor Groove Recognition

To explore more broadly the structural landscape for minor groove recognition, a panel of five-membered aromatic heterocycles was synthesized and incorporated into DNA-binding polyamides (Fig. 3.9) [53, 54]. Because we have found that the Hp residue can degrade over time in the presence of acid or free radicals, a more robust thymine-selective element will be needed for biological applications. Remarkably, none of the heterocycles tested (in the context of an eight-ring hairpin) revealed major advantages over conventional pyrrole-imidazole polyamides. Although ring pairings of Py with either N-methylpyrazole (Pz) or 3-methylthiophene (Tn) are exceptionally selective for A/T bases, displaying no binding to G/C bases in the measured range, incorporation of 4-methylthiazole (Th), furan (Fr), or 3-hydroxythiophene (Ht) abolishes binding entirely. Molecular modeling indicates that relatively small changes to a single heterocycle present in a contiguous four-ring system can induce large alterations in the curvature and electronic properties of the ligand [54]. There appears to be a very narrow window of five-membered ring architectures that are effective for minor groove binding.

The benzimidazole ring system represents a different structural framework, which is amenable to functionalization on the six-membered ring and appears to impart a curvature that is complementary to DNA [55]. Indeed, the classic minor groove-binding Hoechst dyes are composed of benzimidazole units, and a number of derivatives of these molecules have been prepared. We have incorporated benzimidazole derivatives into the backbone of hairpin polyamides in a manner that preserves critical hydrogen bonding contacts and overall molecular shape (Fig. 3.10) [56, 57]. The imidazopyridine (Ip) and hydroxybenzimidazole (Hz) rings are introduced into polyamides as dimeric subunits PyIp and PyHz, respectively, in which the Py ring is directly connected to the benzimidazole derivative without an intervening amide bond [56, 57]. DNase I footprinting indicates that the Ip/Py and Hz/Py pairs are functionally identical, at least in some sequence contexts, to the analogous five-membered ring pairs Im/Py and Hp/Py [56]. Importantly, Hz-containing polyamides are chemically robust, making the Hz/Py pair a strong candidate for replacing Hp/Py in biological studies.

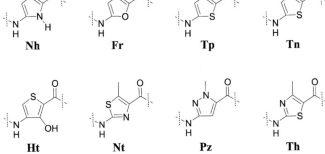

Fig. 3.9 Novel five-membered heterocyclic amino acids that have been incorporated into hairpin polyamides. All residues are shown with the functionality that faces the DNA minor groove towards the bottom-right

3.8
Sequence Specific Alkylation of DNA

The availability of sequence-specific DNA binding molecules led to the development of bifunctional polyamides that covalently react with the minor groove of DNA. Two classes of alkylating agents were conjugated to the hairpin "turn" unit and bound proximal to their alkylation sites by a DNA-binding polyamide [58, 59].

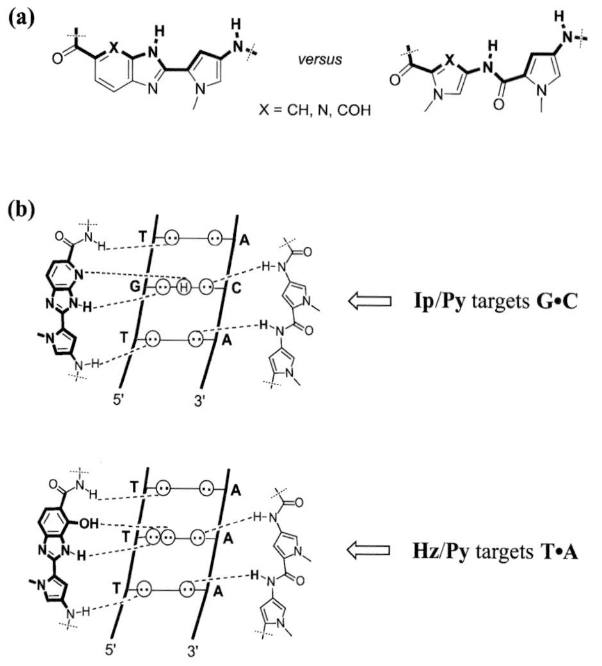

(a)

versus

X = CH, N, COH

(b)

⇐ **Ip/Py targets G•C**

⇐ **Hz/Py targets T•A**

Fig. 3.10 Recognition of the DNA minor groove with benzimidazole derivatives. (a) Structure of the dimeric core for Py-benzimidazole (Bi), Py-hydroxybenzimidazole (Hz) and Py-imidazopyridine (Ip) (left) in comparison with the five-membered ring system (right). H-bonding surfaces along the recognition sites are shown in bold. (b) Postulated hydrogen-bonding models of the 1:1 polyamide:DNA complexes containing the Ip/Py pair (top) targeting G·C and Hz/Py pair (bottom) targeting T·A. Other symbols are defined in Fig. 3.4

Polyamide–*seco*-CBI conjugates showed alkylation at adenines proximal to the binding site within 12 h at nanomolar concentrations [59]. Likewise, polyamide–chlorambucil conjugates specifically alkylate predetermined sites in the minor groove, albeit with slightly lower alkylation at mismatch sites than the CBI counterparts [58]. It is likely that the slower rate of alkylation for the chlorambucil moiety in the minor groove allows for an increased specificity of alkylation to polyamide match sites. This class of sequence-specific DNA-binding alkylators inhibits polymerase elongation during transcription (Fig. 3.11; Gottesfeld, unpublished results).

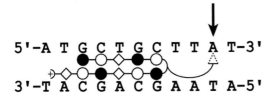

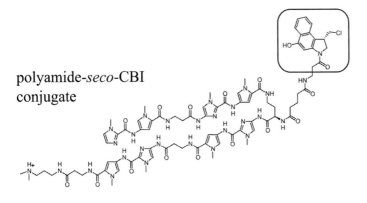

polyamide-*seco*-CBI
conjugate

polyamide-chlorambucil
conjugate

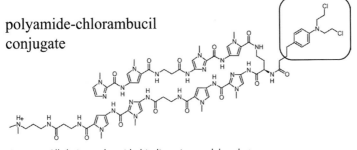

Fig. 3.11 Alkylating polyamide binding site model and structures of polyamide–alkylator conjugates. The dotted triangle represents the alkylating agent. All other symbols are defined in Fig. 3.4. The CBI and chlorambucil alkylator domains are boxed

3.9
Inhibition of Gene Expression

Polyamides bind with high affinity to a wide range of DNA sites, and often competitively displace proteins from DNA. One approach to modifying gene expression involves inhibition of key transcription factor (TF) –DNA complexes in a designated promoter, thus interfering with recruitment of RNA polymerases (Fig. 3.12 and Tab. 3.1) [60]. Significantly, because there are considerably fewer oncogenic TFs than potentially oncogenic signaling proteins, TF inhibition represents a uniquely promising approach to cancer treatment [61]. The transcription factor TFIIIA was chosen as a first target because it regulates a relatively small number of genes and because the contacts between the nine zinc-finger protein and the minor groove had been established. A polyamide bound in the recognition site of TFIIIA suppressed transcription of 5S RNA genes by RNA polymerase

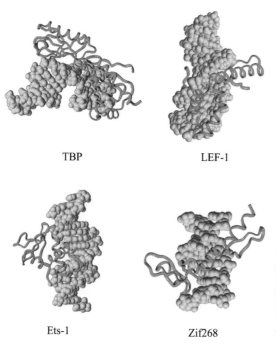

TBP   LEF-1

Ets-1   Zif268

Fig. 3.12 Examples of four different protein–DNA complexes that have been inhibited by polyamides. TBP is a minor groove-binding protein, LEF-1 is an HMG box, Ets-1 is a winged helix-turn-helix, and Zif268 is a zinc finger

Tab. 3.1 DNA-binding proteins that have been inhibited by polyamides. The known DNA binding motifs from NMR or crystal structure data are shown. Significant groove contacts and proposed mechanism of polyamide inhibition are also shown for each protein

Transcription factor	DNA-binding motif	Groove recognition	Proposed mechanism of inhibition
TBP/TFIID	Saddle	Minor	Allosteric
LEF-1	HMG box	Minor	Allosteric
TFIIIB	N/A	Minor	Steric
IE86	Minor	Minor	Steric
Tax	N/A	Minor	Steric
TFIIIA	Cys_2His_2ZnF	Minor/major	Steric
NF-κB	Rel homology domain	Major	Allosteric
LSF	Zinc finger	Major	Allosteric
Ets-1	Winged helix-turn-helix	Major	Phosphate interference
Dpn	Basic helix-loop-helix	Major	Phosphate interference
GCN4	bZip	Major	Phosphate interference

III *in vitro* and in cultured *Xenopus* kidney cells [60]. Further studies used polyamides in combination with recombinant derivatives of TFIIIA subunits to elucidate essential minor groove contacts for the binding of this TF (Tab. 3.1) [62].

Polyamides were then used to target viral genes transcribed by RNA polymerase II (Fig. 3.13). The HIV-1 enhancer/promoter contains binding sites for multiple transcription factors, including TBP, Ets-1, and LEF-1. Two hairpin polyamides designed to bind DNA sequences immediately adjacent to the binding sites for these TFs specifically inhibited binding of each transcription factor and HIV-1 transcription in a cell-free assay [63]. In human blood lymphocytes, treatment with the two polyamides in combination inhibited viral replication by 99%, with no significant decrease in cell viability. Inhibition of viral replication is *indirect* evidence for specific transcription inhibition by polyamides, because other modes of action could be involved, such as modulation of T-cell activation pathways. However, RNase protection assays indicated that the two polyamides did not alter the RNA transcript levels of several cytokine and growth factor genes, suggesting that polyamides do affect transcription directly.

This early biological result spurred a variety of biochemical studies of the interactions of various polyamides with the basal transcription machinery and TF–DNA complexes. Two studies have used promoter scanning to identify sites where polyamide binding inhibits transcription [64, 65]. The method uses a series of DNA constructs with designed polyamide binding sites at varying distances from

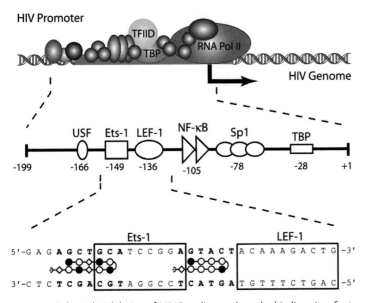

Fig. 3.13 Polyamide inhibition of HIV-1 replication. The HIV promoter is expanded to show the binding sites of crucial transcription factors. The schematic is further expanded to show the binding sites for two polyamides that target the flanking regions of the Ets-1 transcription factor binding region (boxed). All other symbols are defined in Fig. 3.4

the transcription start site. Essential minor groove contacts were identified for a subunit of TFIIIB (possibly TBP) in a *Xenopus* tRNA promoter [64], as well as for TFIID-TFIIA and TBP in the HIV-1 core promoter [65]. The binding of the homo-dimeric basic helix-loop-helix TF Deadpan was investigated using a variant of pro-moter scanning [66]. A series of duplex oligonucleotides based on a *Drosophila* neural promoter were designed, incorporating polyamide binding sites on differ-ent sides of the Deadpan recognition sequence and in different orientations. The TF–DNA complex was inhibited only by a polyamide binding upstream of the homodimer, establishing an asymmetric binding mode for this TF.

The binding of Ets-1 to the HIV-1 enhancer was examined in greater detail, and polyamides were shown to inhibit the formation of a ternary Ets-1–NF-κB–DNA complex [67]. Ets-1 is a winged helix-turn-helix TF, and its key phosphate contacts on either side of the major groove can be disrupted by a polyamide in the adja-cent minor groove. The report provided evidence for cooperative DNA binding by Ets-1 and NF-κB to the HIV-1 enhancer sequence. A different Ets binding site in the HER2/*neu* promoter was targeted with hairpin polyamides that successfully blocked Ets–DNA complex formation and transcription of the HER2/*neu* onco-gene in a cell-free system [68].

Several other protein–DNA interactions have been inhibited with polyamides. In the human T-cell leukemia virus type 1 (HTLV-1) promoter, polyamides tar-geted to G,C-rich regions flanking the viral CRE sites inhibit binding of the Tax protein and Tax transactivation *in vitro* [69]. Bacterial gyrase recognizes a short 5'-GGCC-3' site, and a polyamide targeted to this sequence inhibited gyrase-cata-lyzed strand cleavage at nanomolar concentrations [70]. NF-κB is a TF crucial for development, viral expression, inflammation, and anti-apoptotic responses. The most common form is a p50–p65 heterodimer, which binds DNA in the major groove, making several phosphate contacts throughout the binding site. Polya-mides targeted to the minor groove opposite p50, but not p65, inhibit DNA bind-ing by NF-κB (Fig. 3.14) [71].

Some purely major groove-binding TFs, such as the basic-region leucine zipper (bZIP) protein GCN4, can *co-occupy* the DNA helix in the presence of polyamides [72]. Modified polyamides with an attached Arg-Pro-Arg tripeptide can interfere with major groove-binding proteins by disrupting key phosphate contacts, distort-

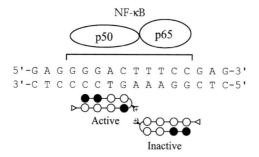

Fig. 3.14 Model for polyamide inhibi-tion of the p65/p50 heterodimeric transcription factor NF-κB. Polyamides designed to target both the p65 and p50 DNA subsites (NF-κB site in bracket) demonstrated that only those targeting the p50 subsite effectively inhibit heterodimer binding. Symbols are defined in Fig. 3.4

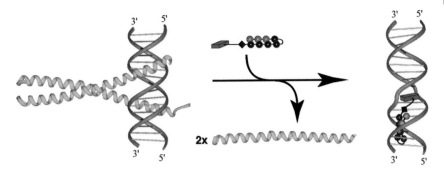

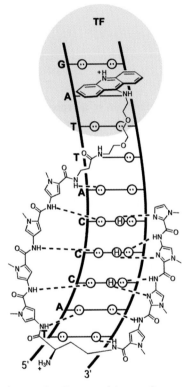

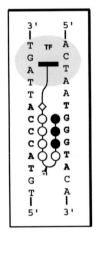

Fig. 3.15 Model for allosteric inhibition of a protein–DNA complex by a polyamide–intercalator conjugate. (Top) The GCN4 homodimer (yellow) is displaced by the intercalating moiety (green) of the polyamide conjugate. Blue and red spheres represent pyrrole and imidazole amino acids, respectively. The blue diamond represents β-alanine. (Bottom, left) Hydrogen-bonding model of an eight-ring hairpin polyamide–intercalator conjugate bound to the minor groove of DNA. The shaded area represents the GCN4 transcription factor (TF). Other symbols are defined in Fig. 3.2. (Bottom, right) Ball-and-stick model of the polyamide–conjugate binding the target site (bold) adjacent to the binding of the protein GCN4. The black bar depicts the acridine intercalator. All other symbols are as defined Fig. 3.4

ing the DNA by charge neutralization, or sterically invading the major groove. An Arg-Pro-Arg–polyamide conjugate successfully inhibited the binding of GCN4 to DNA [72], and further optimization yielded a polyamide derivative with an alkyl diamine substituent that was 10-fold more potent [73]. Polyamide–intercalator conjugates that distort DNA sequence specifically by insertion of an intercalator (e.g. acridine) which unwinds the DNA, are promising candidates for site-selective inhibition of any DNA-binding protein (Fig. 3.15) [74].

In a unique example of altering gene expression by modifying chromatin structure in a complex organism, Laemmli and coworkers targeted satellite regions of *Drosophila* chromosomes with polyamides [75]. Remarkably, polyamides induced specific gain- and loss-of-function phenotypes when fed to developing *Drosophila* embryos.

3.10
Gene Activation

Polyamides can upregulate transcription by inhibition of a repressor protein (derepression) or by recruitment of transcriptional machinery. For example, a hairpin polyamide was shown to block binding of the repressor IE86 to DNA, thereby upregulating transcription of the human cytomegalovirus MIEP [76]. A more complex case involves derepression of the integrated HIV-1 long terminal repeat (LTR). The human protein LSF binds in the promoter region at the LTR and recruits YY1, which then recruits histone deacetylases (HDACs). HDACs subsequently maintain LTR quiescence, which has been implicated in HIV latency, by maintaining a silent stock of pathogen. Three different live cell models demonstrated that polyamides can inhibit LSF binding and increase expression of the integrated HIV-1 promoter (Fig. 3.16) [77]. As with other systems, only polyamides matched to the correct protein binding site induced significant effects. Several existing drug treatments can reduce HIV-1 levels in the blood to below detectable amounts, yet the virus inevitably returns in infected patients. Derepression by inhibition of LSF-DNA binding may eventually allow HIV to be fully eradicated by drug treatments. This approach is particularly promising because LSF is a human protein, which could make the target less susceptible to resistance by HIV-1 mutations.

Recruitment of transcriptional machinery is a fundamentally different approach to gene activation. Polyamides can be thought of as artificial DNA-binding domains that can be linked to an *activation* domain (Fig. 3.17). Such artificial transcription factors have been synthesized and evaluated in cell-free transcription assays [78, 79]. A hairpin polyamide tethered by a 36-atom straight chain linker to the short (20-residue) peptide activation domain AH gives robust activation of transcription, with a size of only 4.2 kDa. Replacing the AH peptide with the shorter yet more potent activator VP2 – derived from the activator domain of the viral activator VP16 – and reducing the linker from 36 to eight atoms provided a

26 TRAUGER, J. W., E. E. BAIRD, and P. B. DERVAN. Recognition of DNA by designed ligands at subnanomolar concentrations. *Nature* **1996**, *382*, 559–561.

27 deCLAIRAC, R. P. L., B. H. GEIERSTANGER, M. MRKSICH, P. B. DERVAN, and D. E. WEMMER. NMR characterization of hairpin polyamide complexes with the minor groove of DNA. *J. Am. Chem. Soc.* **1997**, *119*, 7909–7916.

28 WHITE, S., E. E. BAIRD, and P. B. DERVAN. Orientation preferences of pyrrole-imidazole polyamides in the minor groove of DNA. *J. Am. Chem. Soc.* **1997**, *119*, 8756–8765.

29 HERMAN, D. M., E. E. BAIRD, and P. B. DERVAN. Stereochemical control of the DNA binding affinity, sequence specificity, and orientation preference of chiral hairpin polyamides in the minor groove. *J. Am. Chem. Soc.* **1998**, *120*, 1382–1391.

30 HERMAN, D. M., J. M. TURNER, E. E. BAIRD, and P. B. DERVAN. Cycle polyamide motif for recognition of the minor groove of DNA. *J. Am. Chem. Soc.* **1999**, *121*, 1121–1129.

31 MELANDER, C., D. M. HERMAN, and P. B. DERVAN. Discrimination of A/T sequences in the minor groove of DNA within a cyclic polyamide motif. *Chem.-Eur. J.* **2000**, *6*, 4487–4497.

32 BALIGA, R., E. E. BAIRD, D. M. HERMAN, C. MELANDER, P. B. DERVAN, and D. M. CROTHERS. Kinetic consequences of covalent linkage of DNA binding polyamides. *Biochemistry* **2001**, *40*, 3–8.

33 GREENBERG, W. A., E. E. BAIRD, and P. B. DERVAN. A comparison of H-pin and hairpin polyamide motifs for the recognition of the minor groove of DNA. *Chem.-Eur. J.* **1998**, *4*, 796–805.

34 OLENYUK, B., C. JITIANU, and P. B. DERVAN. Parallel synthesis of H-pin polyamides by alkene metathesis on solid phase. *J. Am. Chem. Soc.* **2003**, *125*, 4741–4751.

35 HECKEL, A. and P. B. DERVAN. U-pin polyamide motif for recognition of the DNA minor groove. *Chem. Eur. J.* **2003**, *9*, 1–14.

36 TURNER, J. M., S. E. SWALLEY, E. E. BAIRD, and P. B. DERVAN. Aliphatic/aromatic amino acid pairings for polyamide recognition in the minor groove of DNA. *J. Am. Chem. Soc.* **1998**, *120*, 6219–6226.

37 WANG, C. C. C., U. ELLERVIK, and P. B. DERVAN. Expanding the recognition of the minor groove of DNA by incorporation of *β*-alanine in hairpin polyamides. *Bioorg. Med. Chem.* **2001**, *9*, 653–657.

38 KELLY, J. J., E. E. BAIRD, and P. B. Dervan. Binding site size limit of the 2:1 pyrrole-imidazole polyamide–DNA motif. *Proc. Natl. Acad. Sci. USA* **1996**, *93*, 6981–6985.

39 SWALLEY, S. E., E. E. BAIRD, and P. B. DERVAN. A pyrrole-imidazole polyamide motif for recognition of eleven base pair sequences in the minor groove of DNA. *Chem.-Eur. J.* **1997**, *3*, 1600–1607.

40 TRAUGER, J. W., E. E. BAIRD, and P. B. DERVAN. Cooperative hairpin dimers for recognition of DNA by pyrrole-imidazole polyamides. *Angew. Chem.-Int. Edit.* **1998**, *37*, 1421–1423.

41 HERMAN, D. M., E. E. BAIRD, and P. B. DERVAN. Tandem hairpin motif for recognition in the minor groove of DNA by pyrrole-imidazole polyamides. *Chem.-Eur. J.* **1999**, *5*, 975–983.

42 KERS, I. and P. B. DERVAN. Search for the optimal linker in tandem hairpin polyamides. *Bioorg. Med. Chem.* **2002**, *10*, 3339–3349.

43 WEYERMANN, P. and P. B. DERVAN. Recognition of ten base pairs of DNA by head-to-head hairpin dimers. *J. Am. Chem. Soc.* **2002**, *124*, 6872–6878.

44 MAESHIMA, K., S. JANSSEN, and U. K. LAEMMLI. Specific targeting of insect and vertebrate telomeres with pyrrole and imidazole polyamides. *EMBO J.* **2001**, *20*, 3218–3228.

45 DERVAN, P. B. and A. R. URBACH. In *Essays in Contemporary Chemistry*, G. QUINKERT, and M. V. KISAKÜREK (Eds), pp. 327–339, Verlag Helv. Chim. Acta, Zürich, **2001**.

46 URBACH, A. R. and P. B. DERVAN. Toward rules for 1:1 polyamide:DNA recognition. *Proc. Natl. Acad. Sci. USA* **2001**, *98*, 4343–4348.

47 URBACH, A. R., J. J. LOVE, S. A. Ross, and P. B. DERVAN. Structure of a *β*-alanine linked polyamide bound to a full helical

turn of purine tract DNA in the 1 : 1 motif. *J. Mol. Biol.* **2002**, *320*, 55–71.

48 BAIRD, E. E. and P. B. DERVAN. Solid phase synthesis of polyamides containing imidazole and pyrrole amino acids. *J. Am. Chem. Soc.* **1996**, *118*, 6141–6146.

49 WURTZ, N. R., J. M. TURNER, E. E. BAIRD, and P. B. DERVAN. Fmoc solid phase synthesis of polyamides containing pyrrole and imidazole amino acids. *Org. Lett.* **2001**, *3*, 1201–1203.

50 FATTORI, D., O. KINZEL, P. INGALLINELLA, E. BIANCHI, and A. PESSI. A practical approach to the synthesis of hairpin polyamide–peptide conjugates through the use of a safety-catch linker. *Bioorg. Med. Chem. Lett.* **2002**, *12*, 1143–1147.

51 SWALLEY, S. E., E. E. BAIRD, and P. B. DERVAN. Effects of γ-turn and β-tail amino acids on sequence-specific recognition of DNA by hairpin polyamides. *J. Am. Chem. Soc.* **1999**, *121*, 1113–1120.

52 BELITSKY, J. M., D. H. NGUYEN, N. R. WURTZ, and P. B. DERVAN. Solid-phase synthesis of DNA binding polyamides on oxime resin. *Bioorg. Med. Chem.* **2002**, *10*, 2767–2774.

53 NGUYEN, D. H., J. W. SZEWCZYK, E. E. BAIRD, and P. B. DERVAN. Alternative heterocycles for DNA recognition: An N-methylpyrazole/N-methylpyrrole pair specifies for A·T/T·A base pairs. *Bioorg. Med. Chem.* **2001**, *9*, 7–17.

54 MARQUES, M. A., R. M. DOSS, A. R. URBACH, and P. B. DERVAN. Toward an understanding of the chemical etiology for DNA minor-groove recognition by polyamides. *Helv. Chim. Acta.* **2002**, *85*, 4485–4517.

55 MINEHAN, T. G., K. GOTTWALD, and P. B. DERVAN. Molecular recognition of DNA by Hoechst benzimidazoles: Exploring beyond the pyrrole-imidazole-hydroxypyrrole polyamide-pairing code. *Helv. Chim. Acta.* **2000**, *83*, 2197–2213 and references cited therein.

56 RENNEBERG, D. and P. B. DERVAN. Imidazopyridine/pyrrole and hydroxybenzimidazole/pyrrole pairs for DNA minor groove recognition. *J. Am. Chem. Soc.* **2003**, *125*, 5707–5716.

57 BRIEHN, C. A., P. WEYERMANN, and P. B. DERVAN. Alternative heterocycles for DNA recognition: The benzimidazole/imidazole pair. *Eur. J. Chem.* **2003**, *9*, 2110–2122.

58 WURTZ, N. R. and P. B. DERVAN. Sequence-specific alkylation of DNA by hairpin pyrrole-imidazole polyamide conjugates. *Chem. Biol.* **2000**, *7*, 153–161.

59 CHANG, A. Y. and P. B. DERVAN. Strand selective cleavage of DNA by diastereomers of hairpin polyamide–*seco*-CBI conjugates. *J. Am. Chem. Soc.* **2000**, *122*, 4856–4864.

60 GOTTESFELD, J. M., L. NEELY, J. W. TRAUGER, E. E. BAIRD, and P. B. DERVAN. Regulation of gene expression by small molecules. *Nature* **1997**, *387*, 202–205.

61 DARNELL, J. E. Transcription factors as targets for cancer therapy. *Nature Rev. Cancer* **2002**, *2*, 740–749.

62 NEELY, L., J. W. TRAUGER, E. E. BAIRD, P. B. DERVAN, and J. M. GOTTESFELD. Importance of minor groove binding zinc fingers within the transcription factor IIIA–DNA complex. *J. Mol. Biol.* **1997**, *274*, 439–445.

63 DICKINSON, L. A., R. J. GULIZIA, J. W. TRAUGER, E. E. BAIRD, D. E. MOSIER, J. M. GOTTESFELD, and P. B. DERVAN. Inhibition of RNA polymerase II transcription in human cells by synthetic DNA-binding ligands. *Proc. Natl. Acad. Sci. USA* **1998**, *95*, 12890–12895.

64 MCBRYANT, S. J., E. E. BAIRD, J. W. TRAUGER, P. B. DERVAN, and J. M. GOTTESFELD. Minor groove DNA–protein contacts upstream of a tRNA gene detected with a synthetic DNA binding ligand. *J. Mol. Biol.* **1999**, *286*, 973–981.

65 EHLEY, J. A., C. MELANDER, D. HERMAN, E. E. BAIRD, H. A. FERGUSON, J. A. GOODRICH, P. B. DERVAN, and J. M. GOTTESFELD. Promoter scanning for transcription inhibition with DNA-binding polyamides. *Mol. Cell. Biol.* **2002**, *22*, 1723–1733.

66 WINSTON, R. L., J. A. EHLEY, E. E. BAIRD, P. B. DERVAN, and J. M. GOTTESFELD. Asymmetric DNA binding by a homodimeric bHLH protein. *Biochemistry* **2000**, *39*, 9092–9098.

67 DICKINSON, L. A., J. W. TRAUGER, E. E. BAIRD, P. B. DERVAN, B. J. GRAVES, and J. M. GOTTESFELD. Inhibition of Ets-1 DNA binding and ternary complex formation between Ets-1, NF-κB, and DNA by a designed DNA-binding ligand. *J. Biol. Chem.* **1999**, *274*, 12765–12773.

68 CHIANG, S. Y., R. W. BURLI, C. C. BENZ, L. GAWRON, G. K. SCOTT, P. B. DERVAN, and T. A. BEERMAN. Targeting the Ets binding site of the HER2/*neu* promoter with pyrrole-imidazole polyamides. *J. Biol. Chem.* **2000**, *275*, 24246–24254.

69 LENZMEIER, B. A., E. E. BAIRD, P. B. DERVAN, and J. K. NYBORG. The Tax protein–DNA interaction is essential for HTLV-I transactivation *in vitro*. *J. Mol. Biol.* **1999**, *291*, 731–744.

70 SIMON, H., L. KITTLER, E. BAIRD, P. DERVAN, and C. ZIMMER. Selective inhibition of DNA gyrase *in vitro* by a GC specific eight-ring hairpin polyamide at nanomolar concentration. *FEBS Lett.* **2000**, *471*, 173–176.

71 WURTZ, N. R., J. L. POMERANTZ, D. BALTIMORE, and P. B. DERVAN. Inhibition of DNA binding by NF-κB with pyrrole-imidazole polyamides. *Biochemistry* **2002**, *41*, 7604–7609.

72 BREMER, R. E., E. E. BAIRD, and P. B. DERVAN. Inhibition of major-groove-binding proteins by pyrrole-imidazole polyamides with an Arg-Pro-Arg positive patch. *Chem. Biol.* **1998**, *5*, 119–133.

73 BREMER, R. E., N. R. WURTZ, J. W. SZEWCZYK, and P. B. DERVAN. Inhibition of major groove DNA binding bZIP proteins by positive patch polyamides. *Bioorg. Med. Chem.* **2001**, *9*, 2093–2103.

74 FECHTER, E. J. and P. B. DERVAN. Allosteric inhibition of protein–DNA complexes by polyamide–intercalator conjugates. *J. Am. Chem. Soc.* **2003**, *125*, 8476–8485.

75 JANSSEN, S., O. CUVIER, M. MULLER, and U. K. LAEMMLI. Specific gain- and loss-of-function phenotypes induced by satellite-specific DNA-binding drugs fed to *Drosophila melanogaster*. *Mol. Cell* **2000**, *6*, 1013–1024.

76 DICKINSON, L. A., J. W. TRAUGER, E. E. BAIRD, P. GHAZAL, P. B. DERVAN, and J. M. GOTTESFELD. Anti-repression of RNA polymerase II transcription by pyrrole-imidazole polyamides. *Biochemistry* **1999**, *38*, 10801–10807.

77 COULL, J. J., G. C. HE, C. MELANDER, V. C. RUCKER, P. B. DERVAN, and D. M. Margo-

lis. Targeted derepression of the human immunodeficiency virus type 1 long terminal repeat by pyrrole-imidazole polyamides. *J. Virol.* **2002**, *76*, 12349–12354.

78 MAPP, A. K., A. Z. ANSARI, M. PTASHNE, and P. B. DERVAN. Activation of gene expression by small molecule transcription factors. *Proc. Natl. Acad. Sci. USA* **2000**, *97*, 3930–3935.

79 ANSARI, A. Z., A. K. MAPP, D. H. NGUYEN, P. B. DERVAN, and M. PTASHNE. Towards a minimal motif for artificial transcriptional activators. *Chem. Biol.* **2001**, *8*, 583–592.

80 ARORA, P. S., A. Z. ANSARI, T. P. BEST, M. PTASHNE, and P. B. DERVAN. Design of artificial transcriptional activators with rigid poly-L-proline linkers. *J. Am. Chem. Soc.* **2002**, *124*, 13067–13071.

81 GOTTESFELD, J. M., C. MELANDER, R. K. SUTO, H. RAVIOL, K. LUGER, and P. B. DERVAN. Sequence-specific recognition of DNA in the nucleosome by pyrrole-imidazole polyamides. *J. Mol. Biol.* **2001**, *309*, 615–629.

82 SUTO, R. K., R. S. EDAYATHUMANGALAM, C. L. WHITE, C. MELANDER, J. M. GOTTESFELD, P. B. DERVAN, and K. LUGER. Crystal structures of nucleosome core particles in complex with minor groove DNA-binding ligands. *J. Mol. Biol.* **2003**, *326*, 371–380.

83 GOTTESFELD, J. M., J. M. BELITSKY, C. MELANDER, P. B. DERVAN, and K. LUGER. Blocking transcription through a nucleosome with synthetic DNA ligands. *J. Mol. Biol.* **2002**, *321*, 249–263.

84 BELITSKY, J. M., S. J. LESLIE, P. S. ARORA, T. A. BEERMAN, and P. B. DERVAN. Cellular uptake of N-methylpyrrole/N-methylimidazole polyamide–dye conjugates. *Bioorg. Med. Chem.* **2002**, *10*, 3313–3318.

85 CROWLEY, K. S., D. P. PHILLION, S. S. WOODARD, B. A. SCHWEITZER, M. SINGH, H. SHABANY, B. BURNETTE, P. HIPPENMEYER, M. HEITMEIER, and J. K. BARKIN. Controlling the intracellular localization of fluorescent polyamide analogs in cultured cells. *Bioorg. Med. Chem. Lett.* **2003**, *13*, 1565–1570.

86 MARINI, N. J., R. BALIGA, M. J. TAYLOR, S. WHITE, P. SIMPSON, L. TSAI, and E. E. BAIRD. DNA-binding hairpin polyamides with antifungal activity. *Chem. Biol.* **2003**, *10*, 635–644.

87 GYGI, M.P., M.D. FERGUSON, H.C. MEF-
FORD, K.P. LUND, C. O'DAY, P. ZHOU, C.
FRIEDMAN, G. VAN DEN ENGH, M.L. STO-
LOWITZ, and B.J. TRASK. Use of fluores-
cent sequence-specific polyamides to dis-
criminate human chromosomes by mi-
croscopy and flow cytometry. *Nucleic
Acids Res.* **2002**, *30*, 2790–2799.

88 RUCKER, V.C., S. FOISTER, C. MELANDER,
and P.B. DERVAN. Sequence-specific fluorescence detection of double strand
DNA. *J. Am. Chem. Soc.* **2003**, *125*, 1195–
1202.

89 SUPEKOVA, L., J.P. PEZACKI, A.I. SU, C.J.
LOWETH, R. RIEDL, B. GEIERSTANGER,
P.G. SCHULTZ, and D.E. WEMMER. Genomic effects of polyamide/DNA interactions on mRNA expression. *Chem. Biol.*
2002, *9*, 821–827.

90 AGAMI, R. RNAi and related mechanisms and their potential use for therapy.
Curr. Opin. Chem. Biol. **2002**, *6*, 829–834.

91 KAMATH, R.S., A.G. FRASER, Y. DONG, G.
POULIN, R. DURBIN, M. GOTTA, A. KANA-
PIN, N. LE BOT, S. MORENO, M. SOHR-
MANN, D.P. WELCHMAN, P. ZIPPERLEN,
and J. AHRINGER. Systematic functional
analysis of the *Caenorhabditis elegans* genome using RNAi. *Nature* **2003**, *421*,
231–237.

92 MALIK, S. and R.G. ROEDER. Transcriptional regulation through mediator-like
coactivators in yeast and metazoan cells.
Trends Biochem. Sci. **2000**, *25*, 277–283.

93 BENTIN, T. and P.E. NIELSEN. Superior
duplex DNA strand invasion by acridine
conjugated peptide nucleic acids. *J. Am.
Chem. Soc.* **2003**, *125*, 6378–6379.

94 BEST, T.B., B.S. EDELSON, N.G. NICKOLS,
and P.B. DERVAN. Nuclear localization of
pyrrole-imidazole polyamide-fluorescein
conjugates in cell culture. *Proc. Natl.
Acad. Sci. USA* **2003**, *100*, 12063–12068.

4
Peptide Nucleic Acid (PNA):
A Pseudo-peptide with DNA-like Properties

Peter E. Nielsen, Uffe Koppelhus, and Frederik Beck

4.1
Introduction

According to the central dogma in molecular biology, nature has selected nucleic acids for storage (DNA primarily) and transfer of genetic information (RNA) in living cells, whereas proteins fulfil the role of carrying out the instructions stored in the genes in the form of enzymes in metabolism and structural scaffolds of the cells. This division of labor is not complete as RNA also has catalytic (ribozymes) and structural roles. However, no examples of proteins as carriers of genetic information* have yet been identified. Thus nature has "chosen" distinctly different classes of chemical compounds for the central biological functions of genetic storage and expression.

However both classes, nucleic acids and proteins, are linear polymers in which the linear (nucleobase or amino acid) sequence encodes the three-dimensional structure and function of the polymer.

Peptide nucleic acids (PNA) are synthetic molecules that are chimeras or hybrids between peptides (or proteins) and nucleic acids [1–5] (Fig. 4.1). In the broadest definition PNAs are composed of "peptide-linked" amino acids bearing nucleobase pendant groups. Indeed, nucleobase amino acids (of unknown function) have been isolated from plants [6, 7] (Fig. 4.2) and similar molecules were also synthesized and oligomerized some 30 years ago in order to obtain DNA-like properties [8, 9]. However, these early attempts were not met with success and it was the general opinion that DNA-mimicking molecules could not be created via this route. Consequently, when antisense technology entered the scientific scene 25 years ago and novel, biologically stable oligonucleotides with high RNA hybridization affinity were needed, virtually all efforts were centered round a "minimal modification concept" of the "ribose-phosphorodiester" nucleic acid backbone [10]. In the mid-1980s the antigene triple helix concept was born as a general principle to sequence specifically, target double-stranded DNA for transcriptional interference with gene expression [11, 12]. However, this otherwise very attractive approach had (and still has) a few shortcomings. Most importantly, target se-

---

* Prions may be an intriguing exception.

Pseudo-peptides in Drug Discovery. Edited by Peter E. Nielsen
Copyright © 2004 Wiley-VCH Verlag GmbH & Co. KGaA, Weinheim
ISBN: 3-527-30633-1

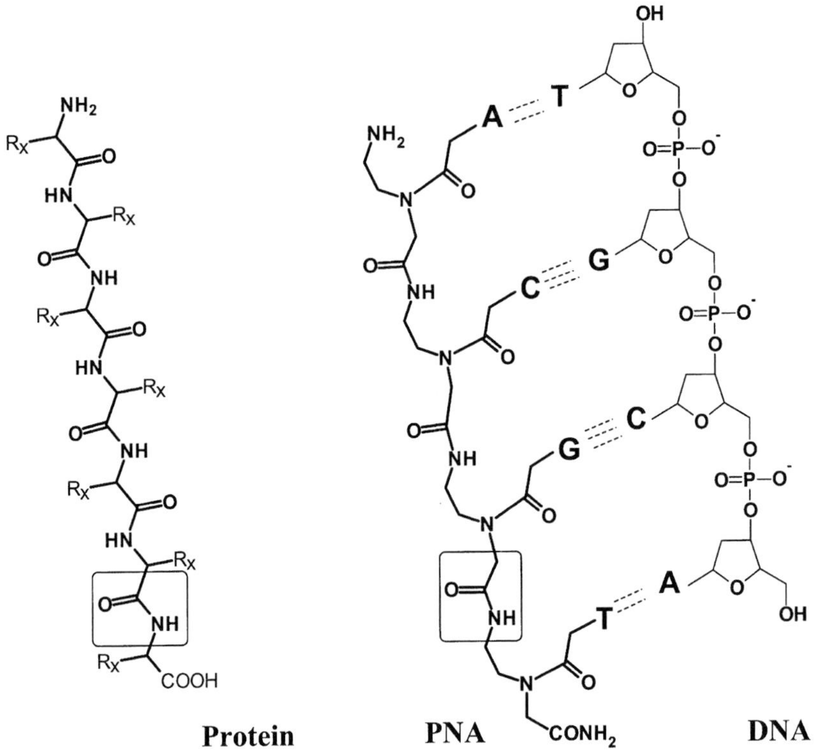

Fig. 4.1 Chemical structure of PNA compared to DNA and protein

quence limitations are imposed due to the requirement for long homopurine/homopyrimidine stretches for triplex recognition. At that time (and this largely still holds true) only adenine and guanines in the target DNA duplex can be targeted by Hoogsteen (or reverse Hoogsteen) base pairing [13]. Furthermore, association of a polyanionic oligonucleotide third strand with a DNA duplex of high negative charge density is thermodynamically unfavored without the presence of relatively high concentrations of (poly)cations.

The development of the aminoethylglycine polyamide (peptide) backbone oligomer with pendant nucleobases linked to the glycine nitrogen via an "acetyl" bridge now often referred to by PNA (peptide nucleic acid; Fig. 4.1), was inspired

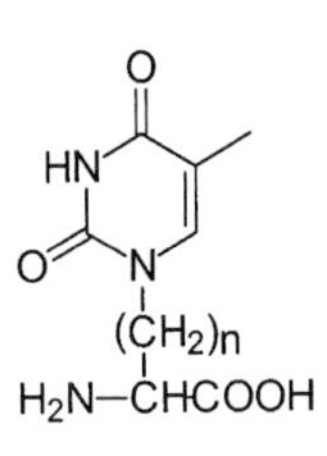

Fig. 4.2 Nucleobase (thymine) amino acid derivative. $n=1$ (9) or $n=4$ (8)

by triple helix targeting of duplex DNA in an effort to combine the recognition power of nucleobases with the versatility and chemical "flexibility" of peptide chemistry [14]. Although these PNAs turned out not to bind sequence complementary targets in duplex DNA very efficiently by conventional triplex formation, they were however, extremely good structural mimics of nucleic acids with a range of interesting properties to be described in this chapter. Furthermore, PNAs formally establish the functional link between peptides and nucleic acids, being a "peptide" with the capacity to carry genetic information.

4.2
Chemistry

The original aminoethylglycine PNA was designed by computer model building to mimic the third strand in a DNA triple helix [1, 14]. By default, it is therefore designed as a DNA structural mimic. Not surprisingly, the number of bonds along the backbone and connecting the backbone with the nucleobase is identical to those of natural nucleic acids (6 + 3). The backbone was also chosen to consist of achiral, single building blocks (N-ethyl glycine and nucleobase acetic acid) which would allow for easy structural change and optimization. Thus the very simple PNA platform has inspired many chemists to explore analogs and derivatives in order to understand and/or improve the properties of this class of DNA mimics (Appendix, Tables A1–A4).

Most modifications have been designed in an effort to stabilize PNA-DNA or PNA-RNA duplexes, but only very few have been met with success in this respect. As the PNA backbone is more flexible (has more degrees of freedom) than the phosphodiester ribose backbone, one could hope that adequate restriction of flexibility would yield higher affinity PNA derivatives. This has led to the synthesis of a large series of conformationally constrained cyclic PNA backbones (Tab. A2). Most of these modifications are detrimental to DNA recognition, but for a few, e.g. a β-alanine-proline (entry 33) and an aminoethylprolyl (entry 40) significant stabilization has been reported. However, no follow-up studies have yet confirmed that such stabilization is general in terms of sequence context and backbone context, i.e. the compatibility of such constrained backbones with e.g. the traditional and much more flexible aeg-PNA backbone. It is therefore too early to judge whether these modifications are generally useful in biological and drug discovery applications. Indeed, one should be very cautious of drawing general conclusions from isolated Tm-stability measurements. However, one derivative, the hypNA-pPNA (entry 57) which is much closer to DNA as it carries a negative charge for every two base pairs, is showing promise in antisense experiments in cell culture [15].

A large variety of nucleobases have also been used and developed in a PNA backbone context (Tab. A4). Several of these are routinely used in PNA applications. The pseudo isocytosine (N1) is employed in the Hoogsteen strand of tri-

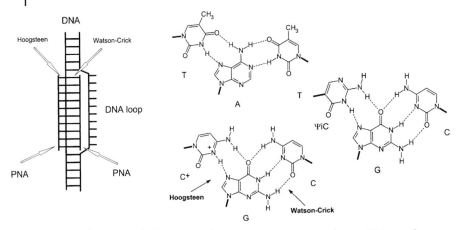

Fig. 4.3 Triplex invasion by homopyrimidine PNA oligomers. One PNA strand binds via Watson–Crick base pairing (preferably in the antiparallel orientation), while the other binds via Hoogsteen base pairing (preferably in the parallel orientation). It is usually advanta-geous to connect the two PNA strands covalently via a flexible linker into a bis-PNA, and to substitute all cytosines in the Hoogsteen strand with pseudoisocytosines (ΨiC), which does not require low pH for N3 "protonation"

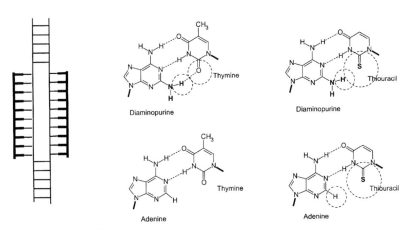

Double Duplex Invasion

Fig. 4.4 Double duplex invasion of pseudo complementary PNAs. In order to obtain efficient binding, the target (and thus the PNAs) should contain at least 50% AT (no other sequence constraints), and in the PNA oligomers all A/T base pairs are substituted with 2,6-diaminopurine/2-thiouracil "base pairs". This base pair is very unstable due to steric hindrance. Therefore the two sequence-complementary PNAs will not be able to bind each other, but they bind their DNA complement very well

plex-forming bis-PNAs for pH-independent recognition of guanine (Fig. 4.3), while the diaminopurine–thiouracil (N2-N4) pair is exploited in double duplex-invading PNAs (Fig. 4.4). Furthermore, diaminopurine, 7-chloro-1,8-naphthyridin-2(1-H)-one (N16) and especially the G-clamp (N32) may find extensive use for stabilization of PNA-DNA (or RNA) duplexes.

4.3
DNA Recognition

Although PNA was originally designed to bind to double-stranded DNA via triple helix recognition in the major groove, the experiments designed to demonstrate such binding most surprisingly revealed a novel binding mode: helix invasion or P-loop formation [1, 16, 17]. Subsequent studies have established triplex invasion of homopurine targets with sequence complementary homopyrimidine PNAs (preferably linked bis-PNAs [18]) as an extremely effective and sequence-specific method of targeting duplex DNA. If the targets are sufficiently long (8–10 bp), the invasion complexes show unprecedented stability ($t_{1/2}$ of the order of days at 37 °C) [19, 20] and under most laboratory (and biological) conditions the complexes are therefore not in equilibrium and consequently under kinetic control. These unique binding characteristics have allowed the development of a number of molecular biology applications that take advantage of high complex stability and/or the P-loop formation, in terms of the extruded single-stranded DNA loop. These applications include PD-loops in which two PNA molecules are employed to create a large DNA loop that can accept a hybridizing oligonucleotide [21] (Fig. 4.5). This complex may in turn initiate DNA replication [22] or serve as a template for circle ligation [23] and subsequent rolling circle amplification [24]. The PNA binding itself blocks the access of DNA-metabolizing enzymes, such as restriction enzymes [25] and methylases and this has been exploited in an Achilles Heel rare cleave technique allowing sequence-specific single cleavage of e.g. yeast chromosomes [26].

In analogy to triplex formation by oligonucleotides, triplex invasion by PNA is limited to homopurine targets. However, in addition the helix invasion process requires opening of the DNA double helix and strand separation. This type of binding is therefore very sensitive to elevated ionic strength. In fact, simple (homopyrimidine) PNAs bind very poorly, if at all, to a duplex DNA target at physiologically relevant ionic strength (e.g. 140 mM K^+/Na^+, 2 mM Mg^{2+}). The invasion process is, however, greatly facilitated by negative DNA supercoiling or by other biological processes that (transiently) unwind the DNA double helix, such as active transcription (e.g. in terms of the transcription bubble) [27, 28]. Thus PNA targeting of duplex DNA by helix invasion may occur much more readily in a living cell than would be predicted from pure *in vitro* data. Furthermore, PNA oligomers conjugated to cationic peptides [29] or 9-aminoacridine (Fig. 4.6) [30] bind their DNA target more efficiently by several orders of magnitude, than naked un-

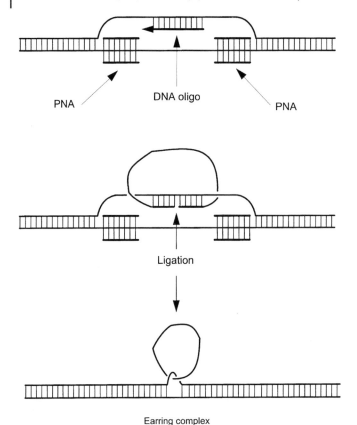

Fig. 4.5 Schematic representation of PD-loop: two homopyrimidine "PNA openers" binding to closely positioned sites create one large DNA loop to which an oligonucleotide can bind. Such a complex may be used to capture the DNA (e.g. via streptavidin beads) by employing a biotinylated oligonucleotide. The complex may be detected by DNA polymerase-mediated extension of the oligonucleotide with radiolabeled precursors. Finally, the complex may be used for template-directed ligation (cyclization) of an oligonucleotide, that may result in an "earring" complex

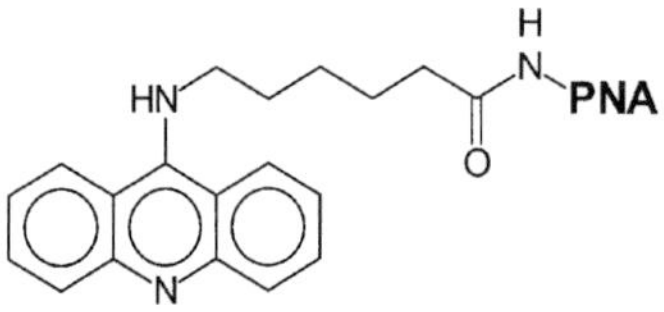

Fig. 4.6 Acridine–PNA conjugate

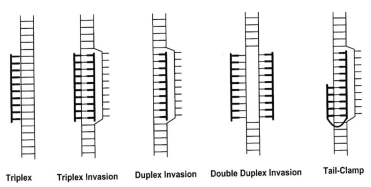

Triplex   **Triplex Invasion**   **Duplex Invasion**   **Double Duplex Invasion**   **Tail-Clamp**

Fig. 4.7 Five different types of PNA–dsDNA complexes. DNA is shown schematically as a ladder, and the PNA oligomers are shown in bold

charged PNA, presumably due to their general DNA affinity that will assure a very high local PNA concentration at the binding sites on the DNA.

PNA targeting of duplex DNA is not limited to homopurine sequences. Under special circumstances (high negative superhelical stress) mixed purine–pyrimidine PNA–peptide conjugates can bind by duplex invasion (Fig. 4.7) [31], but such complexes are of limited stability. However, using a set of pseudo-complementary PNAs containing diaminopurine–thiouracil substitutions, very stable double duplex invasion complexes can be formed (Fig. 4.4) and the only sequence requirement is about 50% AT content. Very recently, it was also demonstrated that reasonably stable helix invasion complexes can be obtained with tail-clamp PNA comprising a short (≥six bases) homopyrimidine bis-PNA clamp and a mixed sequence tail extension [32] (Fig. 4.7).

4.4
Drug Discovery

It was recognized immediately that PNA oligomers have a large potential in drug discovery within the antisense and the antigene fields [1, 14, 33]. This potential has been very slow to be realized, but within the past few years progress has regained strong momentum. This is due to a multitude of developments of which improved delivery systems is a key factor. Also novel molecular biology targets that are particularly sensitive to inhibition by PNA have been identified.

4.4.1
RNA Targeting (Antisense)

Due to the lack of efficient transfection protocols for PNA, the potential of PNA as regulator of gene expression was initially explored mainly in cell-free *in vitro* assays. With respect to translation these studies showed that triplex binding of PNA to mRNA in general constitutes an insurmountable steric barrier to ribosomal action thus resulting in the arrest of initiation or elongation, independent of the position of the target sequence [33–35]. Also, duplex binding PNAs were found to inhibit translation but only if targeted to positions near (or upstream of) the start codon. However, special cases may exist in which translation is efficiently inhibited by duplex-forming PNAs bound to a downstream coding region [36–38]. In addition to translation most other processes involving RNA have been shown in cell-free *in vitro* systems to be vulnerable to sequence-specific PNA interference. Examples are reverse transcription [39–41], telomerase activity [42, 43], primer binding of the HIV-1 genome [44], RNA splicing [45] and ribosomal frameshift [46]. In fact, when properly targeted such processes seems to be considerably more sensitive to PNA-mediated inhibition than translation *per se*. In particular, it has been demonstrated that very low concentrations of PNA efficiently inhibit reverse transcription and telomerase activity [39, 43].

Making use of one or more of the transfection protocols mentioned in a later section of this review, a variety of results have been published claiming intracellular effects of PNA on RNA processing (cf. Tab. 4.1). Some of these studies have focused on translation and include a diverse array of targets spanning from reporter genes (SV40 large T antigene and luciferase [33, 45, 47]) via oncogenes (c-*myc* [48, 49], n-*myc* [50], chimeric promyelocytic leukaemia/retinoic acid receptor-*a* gene [51]) and other genes known or suspected to be involved in human pathogenesis (human microsomal triglyceride transfer protein: atherosclerosis [52]; amyloid precursor protein: Alzheimer's disease [53]; iNOS: autoimmune diseases [54]) to genes coding for proteins of partly unknown function and therefore targeted to obtain more information about their function (oxytocin [55], galanin receptor [56]). Other studies have focused on more specialized RNA processes. These studies include inhibition of HIV replication either by targeting the primer binding site (PBS), the transactivation response element (TAR) [57–59] or the ribosomal frameshift necessary for translation of the HIV-1 *pol* genes [46], inhibition of telomerase activity by targeting the RNA component of the telomerase holoenzyme [43, 60, 61] and disruption of normal or aberrant splicing by targeting intron–exon junctions [45, 62, 63].

Not only eukaryotic cells but also bacteria have successfully been targeted by PNA antisense strategies. Thus it has been shown that PNA complementary to ribosomal RNA or mRNA encoding an essential fatty acid biosynthesis protein, effectively kills *E. coli*. Furthermore, it has been shown that PNA directed to the start codon of the *β*-lactamase gene re-sensitized otherwise resistant *E. coli* to the antibiotic ampicillin [64–66]. Conjugating a simple transporter peptide to the PNA increased the potency significantly, and an even more potent antibacterial PNA

Tab. 4.1 PNA cellular delivery and *ex vivo* effects

PNA	Target	Method	Modification	Cell type/line	Assay	Reference
21-mer	Galanin receptor (ORF)	Direct delivery	Peptide conjugate (penetratin[a]/transportan[b])	Human melanoma Bowes	Receptor activity/protein level (Western blot)	[56]
16-mer	Pre-pro-oxytocin	Direct delivery	Peptide conjugate (retro-inverso penetratin[c])	Primary rat neurons	mRNA level (RT-PCR) Immunocytology	[55]
14-mer (homo-pyrimidine)	Nitric oxide synthase	Direct delivery	PNA peptide conjugate (Phe-Leu)₃	Mouse macrophage RWA264.7	Enzyme activity	[125]
14-mer (homo-pyrimidine)	NO synthetase	Erythrocyte encapsulation	None	Macrophages	NO production and NOS level (Western)	[80]
17-mer	c-myc (ORF-sense)	Direct delivery	NLS peptide[d] conjugate	Burkitt's lymphoma	Protein level (Western blot)/ cell viability	[49]
15-mer	PML-Rar-α AUG)	Cationic liposomes	Adamantyl conjugate	Human lymphocyte (APL) NB4	Protein level (Western blot)/ cell viability	[51]
13-mer	Telomerase (RNA)	Cationic liposomes	PNA/DNA complex	Human prostate cancer DU145	Telomerase activity	[43]
13-mer	Telomerase (RNA)	Cationic liposomes	PNA/DNA complex	Human mammary epithelial (immort.)	Telomerase activity/cell viability/telomere length	[61]
13-mer	Telomerase (RNA)	Electroporation	PNA/DNA complex	AT-SV1, GM05849	Telomerase activity/cell immortality	[60]
11/13-mer	Telomerase (RNA)	Direct delivery	Peptide conjugate (penetratin[c])	JR8/M14, human melanoma	Telomerase activity/cell viability	[101]
13-mer	Telomerase (RNA)	Direct delivery	Tri-lactose conjugate	HepG2 hepatoblastoma	Fluorescence uptake/telomerase activity	[104]
11-mer	None	Direct delivery	Mitochondrial uptake peptide[e]	IMR32, HeLa, a.o.	Only uptake	[138]

Tab. 4.1 (cont.)

PNA	Target	Method	Modification	Cell type/line	Assay	Reference
17-mer	c-myc	Direct delivery	PNA dihydrotestosterone conjugate	Prostatic carcinoma	MYC expression cell viability	[48]
11–18-mer	Luciferase (5'-UTR)	Cationic liposomes	PNA/DNA complex	HeLa	Luciferase activity	[95]
15-mer	IL-5Rα (splice site)	Electroporation	None	BCL₁ lymphoma	RNA synthesis (splicing)	[62]
11-mer	Mithochondrial DNA	Direct delivery	PNA-phosphonium conjugate	143B osteosarcoma/fibroblasts (human)	Biotin uptake/MERRF DNA	[98]
15-mer	HIV-1 gag-pol	Direct delivery	None	H9	Virus production	[46]
16-mer	TAR RNA	Direct delivery	Peptide conjugate (transportan^b))	H9	HIV p24 (ELISA) and luciferase reporter assay	[59]
12–16-mer	TAR RNA	Direct delivery	None	CEM	Luciferase reporter assay	[58]
16-mer	HIV PBS	Direct delivery	None	CEM	Luciferase reporter assay	[57]
13-mer	HIV TAR	Direct delivery	Peptide conjugate (pTat^g))	HeLa	Uptake by fluorescence (endosomal)	[139]
17-mer	NTP/EhErd2 (AUG)	Direct delivery	None	Entamoeba histolytica	Enzyme activity	[68]
18-mer	EGFP (intron)	Electroporation	None/Lys₄	HeLa	GFP synthesis	[45]
14-mer	her2	Direct delivery	Peptide conjugate (penetratin^g)/pTat^g))	HeLa, SKBR3, IMR90, U937, H9, PC19	Uptake by fluorescence (endosomal)	[63]

Tab. 4.1 (cont.)

PNA	Target	Method	Modification	Cell type/line	Assay	Reference
13-mer	None	Direct delivery (nuclear)	Peptide conjugate penetratin[g] + NLS[d]	R3327-AT1 tumor cells	Uptake by fluorescence	[102]
14-mer	MTP	Direct delivery	Tri-GalNAc conjugate	HepG2 liver cells	MTP mRNA (RT-PCR)	[52]
13-mer	bcr/abl	Direct delivery	NLS peptide[d] conjugate	CML cells	Proliferation (+ protein (Western), mRNA (RT PCR))	[63]
12-mer	n-myc	Direct delivery	Somatostatin analog[h] conjugate	IMR32	Proliferation, n-Myc level (Western)	[50]
15-mer	APP	Direct delivery	None	Primary neurons and astrocytes (rat)	Uptake by fluorescence and protein level (Western)	[53]
7-mer bis-PNA	Ribosomal RNA α-sarcin loop	Direct delivery	None	E. coli	Growth inhibition	[64]
10–15-mer	β-lactamase β-glactosidase (AUG)	Direct delivery	None	E. coli	Enzyme activity	[65]
10-mer	acpP (AUG) α-sarcin loop	Direct delivery	Peptide conjugate (KFF[f])	E. coli	Growth inhibition	[66]
Triplex forming bis-PNA		Electroporation	None	Mouse fibroblasts	Mutation induction	[84]
Triplex forming bis-PNA	Globin gene (dsDNA)	Electroporation	None	Monkey kidney CV1	mRNA level (RT-PCR)	[76]

a) Penetratin (pAntp), RQIKIWFQNRRMKWKK;
b) Transportan, GWTLNSAGYLLGKINLAALAKKIL;
c) Retro-inverso penetratin, D-(KKWKMRRNQFWVKVQR);
d) SV40 Nuclear Localization Signal (NLS): PKKKRKV;
e) MSVLTPLLLRGLTGSARRLPVPRAKIHSL;
f) KFFKFFKFFK;
g) pTat, GRKKRRQRRRPPQ;
h) DKDYKDYDK.

was developed by targeting an essential gene involved in fatty acid synthesis, *acp*P. This PNA was shown to inhibit the growth of *E. coli* without affecting human (HeLa) cells [66]. A more detailed study of the mechanism of "KFF"-mediated uptake of PNA in bacteria showed that the peptide facilitated penetration of the outer lipopolysaccharide barrier of Gram-negative bacteria [67].

Unmodified antisense PNA oligomers were also recently shown to downregulate targeted genes in an amoeba (*Entamoeba histolytica*) [68]. These results suggest the potential development of PNA as a new class of antimicrobial drugs.

Finally it should be mentioned that a few *in vivo* (mice) studies have been published in which PNA is employed in an antisense strategy. In one such study PNA is targeted against the translation of the anti-opiod neuropeptide FF (NPFF), and it is claimed that the PNA to a certain degree, was found to abolish the anti-analgesia effect supposedly exerted by NPFF [69]. In another study it was claimed that intravenously-administered antisense PNA targeted to the T cell chemokine receptor CXCR3, downregulated the expression of this receptor whereby the chemokine/cytokine-mediated T cell migration to a skin allograft is reduced, resulting in prolonged survival of the graft [70]. Both these studies raise intriguing perspectives for the therapeutic use of PNA. However, considering the many pitfalls and possible artefacts known to complicate *in vivo* antisense experimentation and interpretation, a series of confirmatory studies is highly warranted, in particular because the lack of adequate controls and molecular biology data is evident in these studies. In contrast, a very convincing *in vivo* study has been carried out by Sazani and colleagues using a transgenic mouse model designed as a positive-readout test for activity, delivery, and distribution of antisense oligomers [47]. In this model, the expressed gene (EGFP-654) encoding enhanced green fluorescent protein (EGFP) is interrupted by an aberrantly spliced mutated intron of the human beta-globin gene. Aberrant splicing of this intron prevents expression of EGFP-654 in all tissues, whereas in tissues and organs that take up a splice site-targeted antisense PNA, correct splicing is restored and EGFP-654 expression upregulated. Sazani and colleagues provide strong evidence that systemically-delivered PNA tagged by four lysine residues, affects gene expression by sequence-specific true antisense activity, thus validating the therapeutic potential of PNA. It is important to note that Sazani and colleagues found that a PNA tagged with only one lysine residue was completely inactive in this very sensitive *in vivo* assay. This fact makes it even more urgent that the results from the other two *in vivo* studies mentioned earlier, are further investigated, as the PNAs employed in these two studies were unmodified.

4.4.2
DNA Targeting (Antigene)

Experiments in cell-free *in vitro* assays have shown that formation of a PNA triplex invasion complex with a homopurine motif on the template strand results in the steric hindrance of RNA polymerases, leading to complete arrest of further mRNA elongation [71]. In contrast to the triplex complexes, double duplex binding of PNA to both the template and the non-template strand by double duplex invasion (Fig. 4.4) does not result in inhibition of transcription, as the RNA polymerase seems capable of overcoming the PNA-DNA binding [72]. An indirect antigene effect has been demonstrated by targeting triplex forming-PNAs to a regulatory motif, a NF-κB transcription factor site, which resulted in a strong inhibition of transcription due to the steric block of NF-κB binding [73]. In 1994 it was shown that P-loop-forming PNAs can induce transcription by *E. coli* RNA polymerase *in vitro* [74]. This interesting finding indicates that dsDNA targets for triplex-invading PNAs are potential promoters with the PNA functioning as an artificial transcription factor and may thus be exploited for experimental and therapeutic purposes. Wang and colleges have recently found that the optimal PNA length for inducing transcription from a 20-bp homopurine stretch was 16–18 nt. The homopurine stretch investigated was cloned in front of two reporter genes (luciferase and green fluorescence protein) and the experiments were carried out either in HeLa nuclear extracts or in human fibroblast, the latter being transfected with the reporter construct after hybridization to the PNA under test [75]. In an earlier study Wang et al. presented data indicating that a P-loop-forming PNA is capable of inducing transcription of an endogenous gene, the developmentally silenced γ-globin gene [76]. This result is somewhat surprising, as triplex invasion of dsDNA *in vitro* is known to be severely impeded by the presence of salt in physiological concentrations. However, it is plausible that transcription or other nuclear processes taking place *in vivo* may render some genomic DNA targets accessible to triplex invasion even at the high concentrations of salt present in the nucleus. Triplex invasion complex forming-PNAs have also been employed in another strategy for artificial gene control. In this strategy a synthetic transcription factor was made by linking a triplex-forming bis-PNA to a 20-residue Gal80-binding peptide as has previously been demonstrated with minor groove binding-polyamides. It was shown that this PNA conjugate was capable of recruiting the Gal 80 protein to a specific DNA site, thus mimicking a fundamental property of native activators, but transcription activation was not studied in this case [77]. Of major interest for the potential antigene application of triplex forming-PNAs is the finding by Corey's group that conjugation of a bis-PNA to a cationic DNA-binding peptide increases the rate of strand invasion by 100-fold relative to the unmodified bis-PNA [29]. Likewise, it was recently shown that simple conjugation of the DNA intercalator 9-aminoacridine, to a PNA (Fig. 4.6) has an analogous effect on dsDNA binding, thus allowing targeting at physiologically relevant ionic strength [30].

A few *ex vivo* and *in vivo* studies have been published claiming an antigene (and antisense) effect of mixed purine/pyrimidine sequence PNA [48, 49, 78–80]. However, as pointed out by us in recent reviews [81, 82] these studies lack fundamental controls such as the inclusion of relevant internal standards as a control for sequence-specific non-antigene/antisense effects, thus confirmatory studies are warranted. The *in vivo* antigene studies from Richelson's group [79, 83] completely lack a rational basis for the claimed effects. First of all there is no evidence that unmodified PNAs should enter cells (either *ex vivo* or *in vivo*). Secondly, based on present knowledge on the binding characteristics of PNA, there is no evidence that short hetero purine/pyrimidine PNA complementary to the template strand of the targeted genes should exert antigene effects such as those persistently claimed by Richelson and coworkers. Given the very indirect biological evidence (body temperature, nociception, blood pressure) that constitutes the basic arguments for an antigene effect in these papers, it is tempting to suspect a confounding biological factor to be the real cause underlying the observed results.

Sequence-targeted gene repair is a most exciting new discipline in drug discovery with the ultimate aim of developing drugs that are capable of repairing disease-causing mutations in somatic cells. Recent *in vitro* and cell culture results have indicated that PNA may also be a tool in this type of application. PNA triplex invasion complexes are somewhat inherently mutagenic [84], but more importantly they may activate the cellular recombination machinery [85], and can also be used as vehicles to target DNA repair domains to the correct genomic site [86]. It will be very interesting to follow this development.

4.4.3
Protein Targeting

Gambari and coworkers have explored the decoy strategy for PNA-mediated intervention of gene expression. In a series of publications they have argued that PNA can be employed in a decoy strategy aimed at reducing the amount of certain "free" transcription factors whereby the transcription of genes from certain promoters will be affected. To demonstrate the validity of this principle they have hybridized short strands of PNA to complementary DNA (PNA/DNA hybrids) or complementary chimeric PNA-DNA-PNA (PDP) strands (PDP/PDP hybrids) and shown that such hybrids can bind NF-κB and Sp1 when the hybrids harbours the binding motifs of these transcription factors. Furthermore, it has been suggested that the presence of such a NF-κB decoy hybrid reduces HIV-1 LTR-driven transcription *in vitro* [87–92]. Although the binding experiments seem convincing and also the indirect downregulation of HIV-1 LTR promoter activity seems to be valid, it is doubtful that PNA, given its radical difference in backbone structure compared to native DNA, is the most suitable DNA analog in such decoy strategies aimed at DNA-binding proteins.

4.4.4
Cellular Delivery

Due to the predominantly hydrophilic nature of PNAs they do not readily cross lipid membranes [93] and enter living cells [94]. Therefore in order to explore the *ex vivo* and *in vivo* potential of PNA as an antisense and/or antigene reagent, a number of different transfection protocols have been devised over recent years.

These protocols comprise a number of techniques involving either unmodified or modified PNA.

Unmodified PNAs have been introduced to cells by microinjection [33], electroporation [60, 62, 76], co-transfection with partially hybridized DNA oligos [43, 61, 95, 96], direct delivery using high concentrations of PNA [46], by the use of streptolycin-O permeabilized cells [84] or by the employment of a mutant bacterial strain having a defective cell wall [64, 65].

To obtain an increased intrinsic capacity to transgress biological membranes, a number of different modifications have been introduced to PNA. These modifications include conjugation of PNA to lipophilic moieties [51, 97, 98], conjugation of PNA to certain so-called cell-penetrating peptides [49, 55, 56, 66, 99–102] and conjugation to different moieties, which are supposed to be internalized by specific cellular receptors [48, 103–105]. The work on cellular delivery of PNA is, like the related work on *ex vivo* and *in vivo* effects of PNA, very difficult to summarize conclusively. First of all, the pronounced diversity of the reporter systems employed makes it impossible to directly compare the studies. Secondly, the widespread use of fluorescence studies in spite of the many inherent pitfalls of this technique makes it sometimes difficult to judge even qualitatively whether a presented result actually indicates cellular uptake. We have recently published a comprehensive review on cellular delivery of PNA [82], with a more detailed assessment of the PNA delivery literature.

A comparative study on the most promising delivery techniques is presently being undertaken in our laboratory. Based on our experience and the preliminary results from this study, we recommend a protocol based on DNA co-transfection to be the first choice for a delivery protocol *ex vivo*.

4.4.5
Pharmacology

Only limited data is so far available on the bioavailability and pharmacokinetics of PNA oligomers. However, it seems safe to conclude that unmodified PNA oligomers appear to exhibit a short serum half-life and limited tissue bioavailability due to rapid excretion through the kidneys [106, 107]. However, chemical modifications such as simple conjugation of cationic peptides (Lys_4, $(KFF)_3$) or backbone modifications, may significantly improve bioavailability [76, 107, 108]. Tissue targeting may also be a viable route. For instance PNA oligomers conjugated to GalNac sugars effectively accumulate in the liver [52, 109]. There have been claims in the literature that PNA oligomers can (readily) cross the blood–brain barrier [79], but this observation has not been supported by independent studies [108, 109].

4.5
Nucleic Acid Detection and Analysis

Although it is outside the scope of this chapter, PNA has also found a number of applications in DNA detection, e.g. in genetic diagnostics [110–112]. In particular, PNA has proven extremely useful as a modifier of PCR amplification by suppressing the amplification (termed PCR clamping) of a wild-type allele over a mutant allele which differs by only one nucleobase [113–118]. By PCR clamping it is possible to amplify a single mutant oncogene from tumor tissue in a natural background of several thousand normal genes [118].

PNA oligomers have also found widespread application as probes for *in situ* hybridization (FISH) both in human diagnostics [119–122] and in environmental detection of microbes [123, 124].

4.6
Nanotechnology

PNA oligomers also constitute (as does DNA) a general molecular recognition system with an almost infinite number of possible codes. Thus PNA molecules may be exploited as "instructed glues" for assembly of other (macro)molecules or nanoparticles. Proof of principle has been obtained by DNA-assisted assembly of gold nanoparticles [125] in addition to the elegant experiments in which three-dimensional "objects" have been built using DNA self-assembly [126, 127]. Only a few examples of research carried out along these lines have so far emerged using PNA. The immobilization of supercoiled DNA on magnetic beads [128] and the (potential) assembly of carbon nanotubes using conjugated PNA oligomers [129] are two such examples. Considering the versatility of PNA chemistry, this area is predicted to flourish in the next decade.

4.7
Pre-RNA World

The hypothesis that our biological world built on the DNA–RNA–protein central dogma was preceded by an RNA world in which RNA molecules carried both the genetic information and executed the gene functions (through ribozyme activity) is now widely accepted [130]. However, it is also well recognized that RNA due to its vulnerability to hydrolysis – especially as a result of catalysis by divalent metal ions – would not have been able to evolve in a harsh pre-biotic environment. Also the formation of RNA under presumed pre-biotic conditions is extremely inefficient. It is not so far-fetched to propose that a peptide nucleic acid-like molecule may have been able to function as a form of pre-biotic genetic material since it

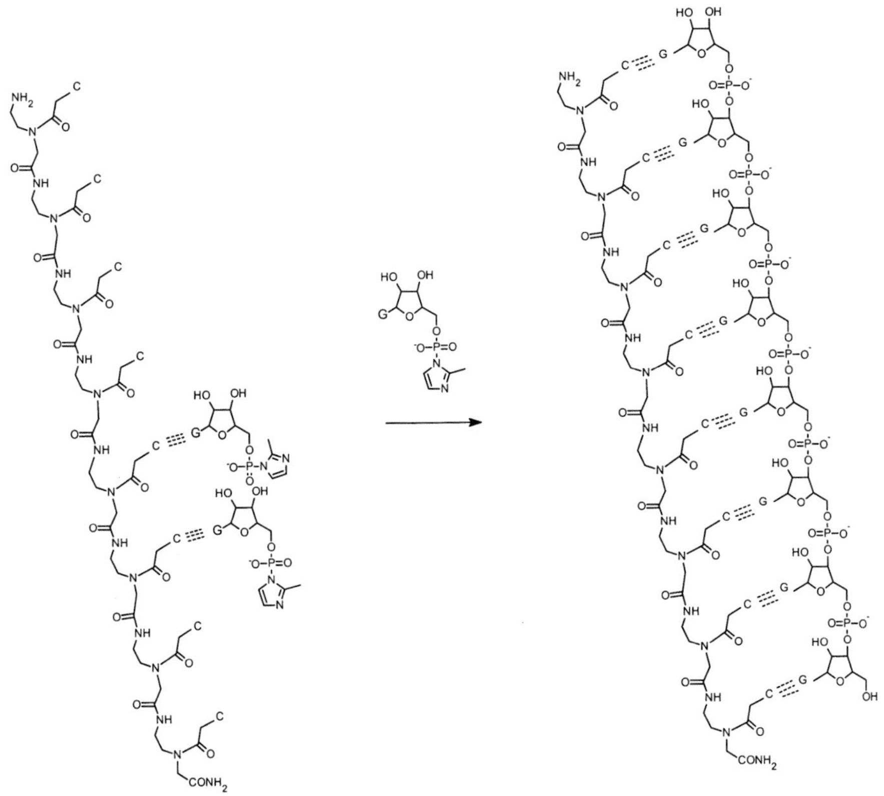

Fig. 4.8 Schematic representation of "chemical replication" of a PNA template to an RNA product

has the required stability [131]. Furthermore, it is conceivable that the very simple chemical structure of PNA may have formed under pre-biotic conditions which were similar to those under which both amino acids [132] and nucleobases [133] have been shown to form quite efficiently. Indeed it was recently found that PNA monomers can form under these conditions [134]. Furthermore, a series of studies have provided proof of principle that sequence information transfer (chemical replication) can take place whereby one PNA oligomer directs the synthesis of a complementary PNA oligomer and similarly it has been shown that a PNA oligomer may direct the chemical oligomerization of a complementary RNA molecule (Fig. 4.8) [135–137]. Obviously such experiments do not by any means provide proof of the existence of a pre-RNA "PNA world", but they do suggest that this might be a possibility. These results may even inspire investigators to search for relicts of nucleobase amino acids in modern cells and organisms. As mentioned earlier, these types of molecules have indeed been identified as metabolites in some plants [6, 7].

4.8
Epilogue

It has been more than 10 years since the introduction of PNA and some of the promises – but of course not all – have been fulfilled. However, new properties and applications have also been discovered and developed along the way. In terms of drug discovery, the road map is still not clear although major progress has been made especially within the last two years. In particular, important new information regarding the behaviour of these molecules in animals is (slowly) emerging and also a range of novel molecular targets have been identified. However, there seems little doubt that cellular delivery and bioavailability is still a major hurdle for which there seems no really effective solution. Nonetheless it is to be hoped that the recent accelerating progress will lead to the possible exploitation of the versatility of PNA in the field of drug development, and there is no doubt that this molecule will keep both chemists and biologist "occupied" for some time to come in the further exploration of its properties, possibilities and applications.

4.9
Acknowledgments

This work has been supported by The Danish Cancer Society, The Lundbeck Foundation and the European Commission (QLK1-CT-2000-01658 and QLRT-2000-00634).

4.10
References

1 NIELSEN P. E., EGHOLM M., BERG R. H., BUCHARDT O. Sequence-selective recognition of DNA by strand displacement with a thymine-substituted polyamide. *Science* 1991, **254**:1497–1500.

2 EGHOLM M., BUCHARDT O., CHRISTENSEN L., BEHRENS C., FREIER S. M., DRIVER D. A., BERG, R. H., KIM, S. K., NORDÉN B., NIELSEN, P. E. PNA hybridizes to complementary oligonucleotides obeying the Watson–Crick hydrogen-bonding rules. *Nature* 1993, **365**:566–568.

3 NIELSEN P. E., HAAIMA G. Peptide nucleic acid (PNA). A DNA mimic with a pseudopeptide backbone. *Chem. Soc. Rev.* 1997, **26**:73–78.

4 NIELSEN P. E. Peptide nucleic acid. A molecule with two identities. *Acc. Chem. Res.* 1999, **32**:624–630.

5 WITTUNG P., NIELSEN P. E., BUCHARDT O., EGHOLM M., NORDÉN B. DNA-like double helix formed by peptide nucleic acid. *Nature* 1994, **368**:561–563.

6 GMELIN R. The free amino acids in the seeds of Acacia willardiana (Mimosaceae). Isolation of willardiin, a new plant amino acid which is probably L-beta-(3-uracil)-alpha-aminopropionic acid. *Hoppe Seylers Z. Physiol. Chem.* 1959; **316**:164–169.

7 BELL E. A., FOSTER R. G. Structure of Lathyrine. *Nature* 1962; **194**:91–92

8 KONING H. DE, Pandit UK Unconventional nucleotide analogues VI. *Rec. Trav. Chim.* 1971; **91**:1069–1080.

9 BUTTREY J. D., JONES A. S., WALKER R. T. Synthetic analogues of polynucleotides XIII. *Tetrahed.* 1975; **31**:73–75.

10 UHLMAN E., PEYMAN, A. Antisense oligonucleotides: A new therapeutic principle. *Chem. Rev.* 1990; **90**:544.

11 MOSER H. E., DERVAN P. B. Sequence-specific cleavage of double helical DNA by triple helix formation. *Science* 1987; **238**:645–650.

12 LE DOAN T., PERROUAULT L., PRASEUTH D., HABHOUB N., DECOUT J. L., THUONG N. T., LHOMME J., HELENE C. Sequence-specific recognition, photocrosslinking and cleavage of the DNA double helix by an oligo-[alpha]-thymidylate covalently linked to an azidoproflavine derivative. *Nucleic Acids Res.* 1987; **15**:7749–7760.

13 Fox K. R. Targeting DNA with triplexes. *Curr. Med. Chem.* 2000; **7**:17–37.

14 NIELSEN P. E., EGHOLM M., BERG R. H., BUCHARDT O. Peptide nucleic acids (PNA): oligonucleotide analogs with a polyamide backbone. In *Antisense Research and Applications*. CROOKE S. T., LEBLEU B. (Eds). CRC Press, Boca Raton, FL; 1993; 364–373.

15 EFIMOV V. A., CHOOB M., BURYAKOVA A. A., PHELAN D., CHAKHMAKHCHEVA O. G. PNA-related oligonucleotide mimics and their evaluation for nucleic acid hybridization studies and analysis. *Nucleosides, Nucleotides Nucleic Acids* 2001; **20**:419–428.

16 CHERNY D. Y., BELOTSERKOVSKII B. P., FRANK-KAMENETSKII M. D., EGHOLM M., BUCHARDT O., BERG R. H., NIELSEN, P. E. DNA unwinding upon strand-displacement binding of a thymine-substituted polyamide to double-stranded DNA. *Proc. Natl Acad. Sci. USA* 1993; **90**:1667–1670

17 NIELSEN P. E., EGHOLM M., BUCHARDT O. Evidence for (PNA)2/DNA triplex structure upon binding of PNA to dsDNA by strand displacement. *J. Mol. Recognit.* 1994; **7**:165–170.

18 EGHOLM M., CHRISTENSEN L., DUEHOLM K. L., BUCHARDT O., COULL J., NIELSEN P. E. Efficient pH-independent sequence-specific DNA binding by pseudoisocyto-sine-containing bis-PNA. *Nucleic Acids Res.* 1995; **23**:217–222.

19 GRIFFITH M. C., RISEN L. M., GREIG M. J., LESNIK E. A., SPRANKLE K. G., GRIFFEY R. H., KIELY J. S., FEIER S. M. Single and bis peptide nucleic acids as triplexing agents: binding and stoichiometry. *J. Am. Chem. Soc.* 1995; **117**:831–832.

20 KOSAGANOV Y. N., STETSENKO D. A., LUBYAKO E. N., KVITKO N. P., LAZURKIN Y. S., NIELSEN P. E. Effect of temperature and ionic strength on the dissociation kinetics and lifetime of PNA-DNA triplexes. *Biochemistry* 2000; **39**:11742–11747.

21 BUKANOV N. O., DEMIDOV V. V., NIELSEN P. E., FRANK-KAMENETSKII M. D. PD-loop: a complex of duplex DNA with an oligonucleotide. *Proc. Natl Acad. Sci. USA* 1998; **95**:5516–5520.

22 DEMIDOV V. V., BROUDE N. E., LAVRENTIEVA-SMOLINA I. V., KUHN H., FRANK-KAMENETSKII M. D. An artificial primosome: design, function, and applications. *ChemBioChem.* 2001; **2**:133–139.

23 KUHN H., DEMIDOV V., FRANK-KAMENETSKII M. D. An earring for the double helix: Assembly of topological links comprising duplex DNA and a circular oligonucleotide. *J. Biomol. Struct. Dyn.* 2000; **2**:221–225.

24 KUHN H., DEMIDOV V. V., FRANK-KAMENETSKII M. D. Rolling-circle amplification under topological constraints. *Nucleic Acids Res.* 2002; **30**:574–580

25 NIELSEN P. E., EGHOLM M., BERG R. H., BUCHARDT O. Sequence specific inhibition of DNA restriction enzyme cleavage by PNA. *Nucleic Acids Res.* 1993; **21**:197–200.

26 VESELKOV A. G., DEMIDOV V. V., NIELSEN P. E., FRANK-KAMENETSKII M. D. A new class of genome rare cutters. *Nucleic Acids Res.* 1996; **24**:2483–2487.

27 LARSEN H. J., NIELSEN P. E. Transcription-mediated binding of peptide nucleic acid (PNA) to double-stranded DNA: sequence-specific suicide transcription. *Nucleic Acids Res.* 1996; **24**:458–463.

28 BENTIN T., NIELSEN P. E. Enhanced peptide nucleic acid binding to supercoiled DNA: possible implications for DNA "breathing" dynamics. *Biochemistry* 1996; **35**:8863–8869.

29 KAIHATSU K., BRAASCH D.A., CANSIZOGLU A., COREY D.R. Enhanced strand invasion by peptide nucleic acid–peptide conjugates. *Biochemistry* 2002; **41**:11118–11125.

30 BENTIN T., NIELSEN P.E. Superior duplex DNA strand invasion by acridine conjugated peptide nucleic acids. *J. Amer. Chem. Soc.* 2003; **125**:6378–6379.

31 ZHANG X., ISHIHARA T., COREY D.R. Strand invasion by mixed base PNAs and a PNA–peptide chimera. *Nucleic Acids Res.* 2000; **28**:3332–3338.

32 BENTIN T., NIELSEN P.E. Combined triplex/duplex invasion of double-stranded DNA by "Tail-Clamp" peptide nucleic acids (PNA). (Submitted)

33 HANVEY J.C., PEFFER N.J., BISI J.E., THOMSON S.A., CADILLA R., JOSEY J.A., RICCA D.J., HASSMAN C.F., BONHAM M.A. Antisense and antigene properties of peptide nucleic acids. *Science* 1992; **258**:1481–1485.

34 KNUDSEN H., NIELSEN P.E. Antisense properties of duplex- and triplex-forming PNAs. *Nucleic Acids Res.* 1996; **24**:494–500.

35 GAMBACORTI-PASSERINI C., MOLOGNI L., BERTAZZOLI C., LE COUTRE P., MARCHESI E., GRIGNANI F., NIELSEN P.E. *In vitro* transcription and translation inhibition by anti-promyelocytic leukemia (PML)/retinoic acid receptor-alpha and anti-PML peptide nucleic acid. *Blood* 1996; **88**:1411–1417.

36 DIAS N., DHEUR S., NIELSEN P.E., GRYAZNOV S., VAN AERSCHOT A., HERDEWIJN P., HÉLÈNE C., SAISON-BEHMOARAS T.E. Antisense PNA tridecamers targeted to the coding region of Ha-ras mRNA arrest polypeptide chain elongation. *J. Mol. Biol.* 1999; **294**:403–416.

37 DIAS N., SÉNAMAUD-BEAUFORT C., LE FORESTIER E., AUVIN C., HÉLÈNE C., SAISON-BEHMOARAS T.E. RNA hairpin invasion and ribosome elongation arrest by mixed base PNA oligomer. *J. Mol. Biol.* 2002; **320**:489–501.

38 MOLOGNI L., LECOUTRE P., NIELSEN P.E., GAMBACORTI-PASSERINI C. Additive antisense effects of different PNAs on the *in vitro* translation of the PML/RAR-alpha gene. *Nucleic Acids Res.* 1998; **26**:1934–1938.

39 KOPPELHUS U., ZACHAR V., NIELSEN P.E., LIU X., EUGEN-OLSEN J., EBBESEN P. EFFICIENT *in vitro* inhibition of HIV-1 gag reverse transcription by peptide nucleic acid (PNA) at minimal ratios of PNA/RNA. *Nucleic Acids Res.* 1997, **25**:2167–2173.

40 LEE R., KAUSHIK N., MODAK M.J., VINAYAK R., PANDEY V.N. Polyamide nucleic acid targeted to the primer binding site of the HIV-1 RNA genome blocks *in vitro* HIV-1 reverse transcription. *Biochemistry* 1998; **37**:900–910.

41 BOULMÉ F., FREUND F., MOREAU S., NIELSEN P.E., GRYAZNOV S., TOULMÉ J.J., LITVAK S. Modified (PNA, 2-O-methyl and phosphoramidate) anti-TAR antisense oligonucleotides as strong and specific inhibitors of *in vitro* HI.V.-1 reverse transcription. *Nucleic Acids Res.* 1998; **26**:5492–5500.

42 NORTON J.C., PIATYSZEK M.A., WRIGHT W.E., SHAY J.W., COREY D.R. Inhibition of human telomerase activity by peptide nucleic acids. *Nature Biotechnol.* 1996; **14**:615–618.

43 HAMILTON S.E., SIMMONS C.G., KATHIRIYA I.S., COREY D.R. Cellular delivery of peptide nucleic acids and inhibition of human telomerase. *Chem. Biol.* 1999; **6**:343–351.

44 KAUSHIK N., TALELE T.T., MONEL R., PALUMBO P., PANDEY V.N. Destabilization of tRNA3Lys from the primer-binding site of HIV-1 genome by anti-A loop polyamide nucleotide analog. *Nucleic Acids Res.* 2001; **29**:5099–5106.

45 SAZANI P., KANG S.H., MAIER M.A., WEI C., DILLMAN J., SUMMERTON J., MANOHARAN M., KOLE R. Nuclear antisense effects of neutral, anionic and cationic oligonucleotide analogs. *Nucleic Acids Res.* 2001; **29**:3965–3974.

46 SEI S., YANG Q.E., O'NEILL D., YOSHIMURA K., MITSUYA H. Identification of a key target sequence to block human immunodeficiency virus type 1 replication within the gag-pol transframe domain. *J. Virol.* 2000; **74**:4621–4633.

47 SAZANI P., GEMIGNANI F., KANG S.H., MAIER M.A., MANOHARAN M., PERSMARK M., BORTNER D., KOLE R. Systemically delivered antisense oligomers upregulate

gene expression in mouse tissues. *Nature Biotechnol.* 2002; **20**:1228–1233.

48 BOFFA L.C., SCARFI S., MARIANI M.R., DAMONTE G, ALLFREY V.G., BENATTI U., MORRIS O.L. Dihydrotestosterone as a selective cellular/nuclear localization vector for anti-gene peptide nucleic acid in prostatic carcinoma cells. *Cancer Res.* 2000; **60**:2258–2262.

49 CUTRONA G., CARPANETO E.M., ULIVI M., RONCELLA S., LANDT O., FERRARINI M., BOFFA L.C. Effects in live cells of a c-*myc* anti-gene PNA linked to a nuclear localization signal. *Nature Biotechnol.* 2000; **18**:300–303.

50 SUN L., FUSELIER J.A., MURPHY W.A., COY D.H. Antisense peptide nucleic acids conjugated to somatostatin analogs and targeted at the n-myc oncogene display enhanced cytotoxity to human neuroblastoma IM.R.32 cells expressing somatostatin receptors. *Peptides* 2002; **23**:1557–1565.

51 MOLOGNI L., MARCHESI E., NIELSEN P.E., GAMBACORTI-PASSERINI C. Inhibition of promyelocytic leukemia (PML)/retinoic acid receptor-alpha and PML expression in acute promyelocytic leukemia cells by anti-PML peptide nucleic acid. *Cancer Res.* 2001; **61**:5468–5473.

52 BIESSEN E.A.L., SLIEDREGT-BOL K., HOEN P.A.C., PRINCE P., VAN DER BILT E., VALENTIJN A.R., MEEUWENOORD N.J., PRINCEN H., BIJSTERBOSCH M.K., VAN DER MAREL G.A., VAN BOOM J.H., VAN BERKEL T.J.C. Design of a targeted peptide nucleic acid prodrug to inhibit hepatic human microsomal triglyceride transfer protein expression in hepatocytes. *Bioconjug. Chem.* 2002; **13**:295–302.

53 ADLERZ L., SOOMETS U., HOLMLUND L., VIIRLAID S., LANGEL Ü., IVERFELDT K. Down-regulation of amyloid precursor protein by peptide nucleic acid oligomer in cultured rat primary neurons and astrocytes. *Neurosci. Lett.* 2003; **336**:55–59.

54 CHIARANTINI L., CERASI A., FRATERNALE A., ANDREONI F., SCARI S., GIOVINE M., CLAVARINO E., MAGNANI M. Inhibition of macrophage iNOS by selective targeting of antisense PNA. *Biochemistry* 2002; **41**:8471–8477.

55 ALDRIAN-HERRADA G., DESARMÉNIEN M.G., ORCEL H., BOISSIN-AGASSE L., MÉRY J., BRUGIDOU J., RABIE A. A peptide nucleic acid (PNA) is more rapidly internalized in cultured neurons when coupled to a retro-inverso delivery peptide. The antisense activity depresses the target mRNA and protein in magnocellular oxytocin neurons. *Nucleic Acids Res.* 1998; **26**:4910–4916.

56 POOGA M., SOOMETS U., HÄLLBRINK M., VALKNA A., SAAR K., REZAEI K., KAHL U., HAO J.X., XU X.J., WISENFELD-HALLIN Z., HÖKFELT T., BARTFAI T., LANGEL Ü. Cell penetrating PNA constructs regulate galanin receptor levels and modify pain transmission *in vivo*. *Nature Biotechnol.* 1998; **16**:857–861.

57 KAUSHIK N., PANDEY V.N. PNA targeting the PBS and A-loop sequences of HIV-1 genome destabilizes packaged tRNA3(Lys) in the virions and inhibits HIV-1 replication. *Virology* 2002; **303**:297–308.

58 KAUSHIK N., BASU A., PANDEY V.N. Inhibition of HIV-1 replication by anti-trans-activation responsive polyamide nucleotide analog. *Antiviral Res.* 2002; **56**:13–27.

59 KAUSHIK N., BASU A., PALUMBO P., MYERS R.L., PANDEY V.N. Anti-TAR polyamide nucleotide analog conjugated with a membrane-permeating peptide inhibits human immunodeficiency virus type 1 production. *J. Virol.* 2002; **76**:3881–3891.

60 SHAMMAS M.A., SIMMONS C.G., COREY D.R., REIS R.J.S. Telomerase inhibition by peptide nucleic acids reverses "immortality" of transformed human cells. *Oncogene* 1999; **18**:6191–6200

61 HERBERT B.S., PITTS A.E., BAKER S.I., HAMILTON S.E., WRIGHT W.E., SHAY J.W., COREY D.R. Inhibition of human telomerase in immortal human cells leads to progressive telomere shortening and cell death. *Proc. Natl Acad. Sci. USA* 1999; **96**:14276–14281.

62 KARRAS J.G., MAIER M.A., LU T., WATT A., MANOHARAN M. Peptide nucleic acids are potent modulators of endogenous pre-mRNA splicing of the murine interleukin-5 receptor-alpha chain. *Biochemistry* 2001; **40**:7853–7859.

63 RAPOZZI V., BURM B. E. A., COGOI S., VAN
DER MAREL G. A., VAN BOOM J. H., QUAD-
RIFOGLIO F., XODO L. E. Antiproliferative
effect in chronic myeloid leukaemia cells
by antisense peptide nucleic acids. *Nucleic Acids Res.* 2002; 30:3712–3721.

64 GOOD L., NIELSEN P. E. Inhibition of
translation and bacterial growth by peptide nucleic acid targeted to ribosomal
RNA. *Proc. Natl Acad. Sci. USA* 1998;
95:2073–2076.

65 GOOD L., NIELSEN P. E. Antisense inhibition of gene expression in bacteria by
PNA targeted to mRNA. *Nature Biotechnol.* 1998; 16:355–358.

66 GOOD L., AWASTHI S. K., DRYSELIUS R.,
LARSSON O., NIELSEN P. E. Bactericidal
antisense effects of peptide–PNA conjugates. *Nature Biotechnol.* 2001; 19:360–
364.

67 ERIKSSON M., NIELSEN P. E., GOOD L.
Cell permeabilization and uptake of antisense peptide–peptide nucleic acid (PNA)
into *Escherichia coli. J. Biol. Chem.* 2002;
277:7144–7147.

68 STOCK R. P., OLVERA A., SANCHEZ R.,
SARALEGUI A., SCARFI S., SANCHEZ-LOPEZ
R., RAMOS M. A., BOFFA L. C., BENATTI
U., ALAGON A. Inhibition of gene expression in *Entamoeba histolytica* with antisense peptide nucleic acid oligomers. *Nature Biotechnol.* 2001; 19:231–234.

69 BONNARD E., MAZARGUIL H., ZAJAC J. M.
Peptide nucleic acids targeted to the
mouse proNPFF(A) reveal an endogenous opioid tonus. *Peptides* 2002;
23:1107–1113.

70 JIANKUO M., XINGBING W., BAOJUN H.,
XIONGWIN W., ZHUOYA L., PING X.,
YONG X., ANTING L., CHUNSONG H.,
FEILI G., JINQUANT T. Peptide Nucleic
Acid antisense prolongs skin allograft
survival by means of blockade of CXCR3
expression directing T cells into graft.
J. Immunol. 2003; 170:1556–1565.

71 NIELSEN P. E., EGHOLM M., BUCHARDT O.
Sequence-specific transcription arrest by
peptide nucleic acid bound to the DNA
template strand. *Gene* 1994; 149:139–145.

72 LOHSE J., DAHL O., NIELSEN P. E. Double
duplex invasion by peptide nucleic acid:
a general principle for sequence-specific

targeting of double-stranded DNA. *Proc.
Natl Acad. Sci. USA* 1999; 96:11804–11808.

73 VICKERS T. A., GRIFFITY M. C., RAMASAMY
K, RISEN L. M., FREIER S. M. Inhibition
of NF-κB specific transcriptional activation by PNA strand invasion. *Nucleic
Acids Res.* 1995; 23:3003–3008.

74 MØLLEGAARD N. E., BUCHARDT O.,
EGHOLM M., NIELSEN P. E. Peptide nucleic acid–DNA strand displacement
loops as artificial transcription promoters. *Proc. Natl Acad. Sci. USA* 1994;
91:3892–3895.

75 WANG G., JING K., BALCZON R., XU X.
Defining the peptide nucleic acids (PNA)
length requirement for PNA binding-in-
duced transcription and gene expression.
J. Mol. Biol. 2001, 313:933–940

76 WANG G., XU X., PACE B., DEAN D. A.,
GLAZER P. M., CHAN P., GOODMAN S. R.,
SHOKOLENKO I. Peptide nucleic acid
(PNA) binding-mediated induction of
human γ-globin gene expression. *Nucleic
Acids Res.* 1999, 27:2806–2813.

77 LIU B., HAN Y., COREY D. R., KODADEK T.
Toward synthetic transcription activators:
recruitment of transcription factors to
DNA by a PNA–peptide chimera. *J. Am.
Chem. Soc.* 2002; 124:1838–1839.

78 TYLER B. M., MCCORMICK D. J., HOSHALL
C. V., DOUGLAS C. L., JANSEN K., LACY
B. W., CUSACK B., RICHELSON E. Specific
gene blockade shows that peptide nucleic
acids readily enter neuronal cells *in vivo.*
FEBS Lett. 2000; 421:280–284.

79 TYLER B. M., JANSEN K., MCCORMICK
D. J., DOUGLAS C. L., BOULES M., STEWART
J. A., ZHAO L., LACY B., CUSACK B., FAUQ
A., RICHELSON E. Peptide nucleic acids
targeted to the neurotensin receptor and
administered i.p. cross the blood–brain
barrier and specifically reduce gene expression. *Proc. Natl Acad. Sci. USA* 1999;
96:7053–7058.

80 SCARFI S., GIOVINE M., GASPARINI A.,
DAMONTE G., MILLO E., POZZOLINI M.,
BENATTI U. Modified peptide nucleic
acids are internalized in mouse macro-
phages RAW 264.7 and inhibit inducible
nitric oxide synthase. *FEB. S. Lett.* 1999;
451:264–268.

81 KOPPELHUS U., NIELSEN P. E. Antisense
properties of peptide nucleic acid (PNA).

In *Basic Principles of Antisense Technology*, CROOKE S. T. (Ed.). 2001, Marcel Dekker, Inc.: New York, pp. 359–374.

82 KOPPELHUS U., NIELSEN P. E. Cellular delivery of peptide nucleic acid (PNA). *Adv. Drug Deliv. Rev.* 2003; **55**:267–280

83 MCMAHON B. M., STEWART J. A., BITNER M. D., FAUQ A., MCCORMICK D. J., RICHELSON E. Peptide nucleic acids specifically cause antigene effects in vivo by systemic injection. *Life Sci.* 2002; **71**:325–337.

84 FARUQI A. F., EGHOLM M., GLAZER P. M. Peptide nucleic acid-targeted mutagenesis of a chromosomal gene in mouse cells. *Proc. Natl Acad. Sci. USA* 1998; **95**:1398–1403.

85 BELOTSERKOVSKII B. P., ZARLING D. A. Peptide nucleic acid (PNA) facilitates multistranded hybrid formation between linear double-stranded DNA targets and RecA protein-coated complementary single-stranded DNA probes. *Biochemistry* 2002; **41**:3686–3692.

86 ROGERS F. A., VASQUEZ K. M., EGHOLM M., GLAZER P. M. Site-directed recombination via bifunctional PNA–DNA conjugates. *Proc. Natl. Acad. Sci. USA* 2002; **99**:16695–16700

87 BORGATTI M., ROMANELLI A., SAVIANO M., PEDONE C., LAMPRONTI I., BREDA L., NASTRUZZI C., BIANCHI N., MISCHIATI C., GAMBARI R. Resistance of decoy PNA–DNA chimeras to enzymatic degradation in cellular extracts and serum. *Oncol. Res.* 2003; **13**:279–287.

88 BORGATTI M., LAMPRONTI I., ROMANELLI A., PEDONE C., SAVIANO M., BIANCHI N., MISCHIATI C, GAMBARI R. Transcription factor decoy molecules based on a peptide nucleic acid (PNA)–DNA chimera mimicking Sp1 binding sites. *J. Biol. Chem.* 2003; **278**:7500–7509.

89 BORGATTI M., BREDA L., CORTESI R., NASTRUZZI C., ROMANELLI A., SAVIANO M., BIANCHI N., MISCHIATIC C, GAMBARI R. Cationic liposomes as delivery systems for double-stranded PNA–DNA chimeras exhibiting decoy activity against NF-κB transcription factors. *Biochem. Pharmacol.* 2002; **64**:609–616.

90 MISCHIATI C., BORGATTI M., FERIOTTO G., RUTIGLIANO C., BREDA L., BIANCHI N., GAMBARI R. Inhibition of hiv-1 ltr-driven *in vitro* transcription by molecular hybrids based on peptide nucleic acids mimicking the NF-κB binding site. *Int. J. Mol. Med.* 2002; **9**:633–639.

91 ROMANELLI A., PEDONE C., SAVIANO M., BIANCHI N., BORGATTI M., MISCHIATI C., GAMBARI R. Molecular interactions between nuclear factor B (NF-κB) transcription factors and a PNADNA chimera mimicking NF-κB binding sites. *Eur. J. Biochem.* 2001; **268**:6066–6075.

92 NASTRUZZI C., CORTESI R., ESPOSITO E., GAMBARI R., BORGATTI M., BIANCHI N., FERIOTTO G., MISCHIATI C. Liposomes as carriers for DNA–PNA hybrids. *J. Controlled Rel.* 2000; **68**:237–249.

93 WITTUNG P., KAJANUS J., EDWARDS K., NIELSEN P., NORDÉN B., MALMSTRÖM B. G. Phospholipid membrane permeability of peptide nucleic acid. *FEB. S. Lett.* 1995; **365**:27–29.

94 BONHAM M. A., BROWN S., BOYD A. L., BROWN P. H., BRUCKENSTEIN D. A., HANVEY J. C., et al. An assessment of the antisense properties of RNase H-competent and steric-blocking oligomers. *Nucleic Acids Res.* 1995; **23**:1197–1203.

95 DOYLE D. F., BRAASCH D. A., SIMMONS C. G., JANOWSKI B. A., COREY D. R. Inhibition of gene expression inside cells by peptide nucleic acids: effect of mRNA target sequence, mismatched bases, and PNA length. *Biochemistry* 2001; **40**:53–64.

96 YUAN X., MA Z., ZHOU W., NIIDOME T., ALBER S., HUANG L., WATKINS S., LI S. Lipid-mediated delivery of peptide nucleic acids to pulmonary endothelium. *Biochem. Biophys. Res. Comm.* 2003; **302**:6–11.

97 LJUNGSTRØM T., KNUDSEN H., NIELSEN P. E. Cellular uptake of adamantyl conjugated peptide nucleic acids. *Bioconjug. Chem.* 1999; **10**:965–972.

98 MURATOVSKA A., LIGHTOWLERS R. N., TAYLOR R. W., TURNBULL D. M., SMITH R. A. J., WILCE J. A., MARTIN S. W., MURPHY M. P. Targeting peptide nucleic acid (PNA) oligomers to mitochondria within cells by conjugation to lipophilic cations: implications for mitochondrial DNA replication, expression and disease. *Nucleic Acids Res.* 2001; **29**:1852–1863.

99 SIMMONS C. G., PITTS A. E., MAYFIELD L. D., SHAY J. W., COREY D. R. Synthesis and membrane permeability of PNA-peptide conjugates. *Bioorg. Med. Chem. Lett.* 1997; 7:3001–3006.

100 KOPPELHUS U., AWASTHI S. K., ZACHAR V., HOLST H. U., EBBESEN P., NIELSEN P. E. Cell-dependent differential cellular uptake of PNA, peptides, and PNA–peptide conjugates. *Antisense Nucleic Acid Drug Dev.* 2002; 12:51–63.

101 VILLA R., FOLINI M., LUALDI S., VERONESE S., DAIDONE M. G., ZAFFARONI N. Inhibition of telomerase activity by a cell-penetrating peptide nucleic acid construct in human melanoma cells. *FEB. S. Lett.* 2000; 473:241–248.

102 BRAUN K., PESCHKE P., PIPKORN R., LAMPEL S., WACHSMUTH M., WALDECK W., FRIEDRICH E., DEBUS J. A biological transporter for the delivery of peptide nucleic acids (PNAs) to the nuclear compartment of living cells. *J. Mol. Biol.* 2002; 318:237–243.

103 BASU S., WICKSTROM E. Synthesis and characterization of a peptide nucleic acid conjugated to a D-peptide analog of insulin-like growth factor 1 for increased cellular uptake. *Bioconjug. Chem.* 1997; 8:481–488.

104 ZHANG X., SIMMONS C. G., COREY D. R. Liver cell specific targeting of peptide nucleic acid oligomers. *Bioorg. Med. Chem. Lett.* 2001; 11:1269–1272.

105 REBUFFAT A. G., NAWROCKI A. R., NIELSEN P. E., BERNASCONI A. G., BERNALMENDEZ E., FREY B. M., FREY F. J. Gene delivery by a steroid–peptide nucleic acid conjugate. *F. A. S. E. B J.* 2002; 16:1426–1428.

106 MCMAHON B. M., MAYS D., LIPSKY J., STEWART J. A., FAUQ A., RICHELSON E. Pharmacokinetics and tissue distribution of a peptide nucleic acid after intravenous administration. *Antisense Nucleic Acid Drug Dev.* 2002; 12:65–70

107 KRISTENSEN E. *In vitro* and *in vivo* studies on pharmacokinetics and metabolism of PNA constructs in rodents. In *Peptide Nucleic Acids: Methods and Protocols*, NIELSEN P. E. (Ed.). 2002, Humana Press (Totowa, N. J., United States): Copenhagen, pp. 259–269.

108 DOLLE F., BOISGARD R., HINNEN F., HAMZAVI R., NIELSEN P. E., TAVITIAN B. (in preparation).

109 HAMZAWI R., DOLLE F., TAVITIAN R., DAHL O., NIELSEN P. E. (Submitted).

110 NIELSEN P. E. Peptide nucleic acid: a versatile tool in genetic diagnostics and molecular biology. *Curr. Opin. Biotechnol.* 2001; 12:16–20

111 DEMIDOV V. V. PD-loop technology: PNA openers at work. *Expert. Rev. Mol. Diagn.* 2001, 1:343–351.

112 DEMIDOV V. V. New kids on the block: emerging PNA-based DNA diagnostics. *Expert. Rev. Mol. Diagn.* 2002, 2:199–201.

113 ØRUM H., NIELSEN P. E., EGHOLM M., BERG R. H., BUCHARDT O., STANLEY C. Single base pair mutation analysis by PNA directed PCR clamping. *Nucleic Acids Res.* 1993; 21:5332–5336.

114 BEHN M., SCHUERMANN M. Sensitive detection of *p53* gene mutations by a "mutant-enriched" PCR-SSCP technique. *Nucleic Acids Res.* 1998; 26:1356–1358.

115 MURDOCK D. G., CHRISTACOS N. C., WALLACE D. C. The age-related accumulation of a mitochondrial DNA control region mutation in muscle, but not brain, detected by a sensitive PNA-directed PCR clamping based method. *Nucleic Acids Res.* 2000; 28:4350–4355.

116 MYAL Y., BLANCHARD A., WATSON P., CORRIN M., SHIU R., IWASIOW B. Detection of genetic point mutations by peptide nucleic acid-mediated polymerase chain reaction clamping using paraffin-embedded specimens. *Anal. Biochem.* 2000; 285:169–172.

117 VON WINTZINGERODE F., LANDT O., EHRLICH A., GOBEL U. B. Peptide nucleic acid-mediated PCR clamping as a useful supplement in the determination of microbial diversity. *Appl. Environ. Microbiol.* 2000; 66:549–557.

118 BEHN M., THIEDE C., NEUBAUER A., PANKOW W., SCHUERMANN M. Facilitated detection of oncogene mutations from exfoliated tissue material by a PNA-mediated 'enriched PCR' protocol. *J. Pathol.* 2000; 190:69–75.

119 LANSDORP P. M., VERWOERD N. P., VAN DE RIJKE F. M., DRAGOWSKA V., LITTLE M.-T., DIRKS R. W., RAAP A. K., TANKE H. J. Heterogeneity in telomere length of human chromosomes. *Hum. Mol. Genet.* 1996; 5:685–691.

120 CHEN C., WU B., WIE T., EGHOLM M., STRAUSS W. M. Unique chromosome identification and sequence-specific structural analysis with short PNA oligomers. *Mamm. Genome* 2000; 11:384–391.

121 HONGMANEE P., STENDER H., RASMUSSEN O. F. Evaluation of a fluorescence *in situ* hybridization assay for differentiation between tuberculous and nontuberculous *Mycobacterium* species in smears of Lowenstein–Jensen and mycobacteria growth indicator tube cultures using peptide nucleic acid probes. *J. Clin. Microbiol.* 2001; 39:1032–1035.

122 DROBNIEWSKI F. A., MORE P. G., HARRIS G. S. Differentiation of *Mycobacterium tuberculosis* complex and nontuberculous mycobacterial liquid cultures by using peptide nucleic acid-fluorescence *in situ* hybridization probes. *J. Clin. Microbiol.* 2000; 38:444–447.

123 PERRY-O'KEEFE H., STENDER H., BROOMER A., OLIVEIRA K., COULL J., HYLDIG-NIELSEN J. J. Filter-based PNA *in situ* hybridization for rapid detection, identification and enumeration of specific microorganisms. *J. Appl. Microbiol.* 2001; 90:180–189.

124 STENDER H., SAGE A., OLIVEIRA K., BROOMER A. J., YOUNG B., COULL J. Combination of ATP-bioluminescence and PNA probes allows rapid total counts and identification of specific microorganisms in mixed populations. *J. Microbiol. Methods* 2001; 46:69–75.

125 MIRKIN C. A., LETSINGER R. L., MUCIC R. C., STORHOFF J. J. A DNA-based method for rationally assembling nanoparticles into macroscopic materials. *Nature* 1996; 382:607–609.

126 SEEMAN N. C. DNA in a material world. *Nature* 2003; 421:427–431.

127 SEEMAN N. C. DNA engineering and its application to nanotechnology. *Trends Biotechnol.* 17:437–443.

128 BENTIN T., NIELSEN P. E. *In vitro* transcription of a torsionally constrained template. *Nucleic Acids Res.* 2002; 30:803–809.

129 WILLIAMS K. A., VEENHUIZEN P. T., DE LA TORRE B. G., ERITJA R., DEKKER C. Nano-

technology: carbon nanotubes with DNA recognition. *Nature* 2002; 420:761.

130 JOYCE G. F. The antiquity of RNA-based evolution. *Nature* 2002; 418:214–221.

131 NIELSEN P. E. Peptide nucleic acid (PNA): A model structure for the primordial genetic material. *Orig. Life Evolut. Biosphere* 1993; 23:323–327.

132 MILLER S. L. A production of amino acids under possible primitive earth conditions. *Science* 1953; 117:528–529.

133 ORO J. Synthesis of adenine from ammonium cyanide. *Biochem. Biophys. Res. Commun.* 1960; 2:407–412.

134 NELSON K. E., LEVY M., MILLER S. L. Peptide nucleic acids rather than RNA may have been the first genetic molecule. *Proc. Natl Acad. Sci. USA* 2000; 97:3868–3871.

135 BÖHLER C., NIELSEN P. E., ORGEL L. E. Template switching between PNA and RNA oligonucleotides. *Nature* 1995; 376:578–581.

136 SCHMIDT J. G., NIELSEN P. E., ORGEL L. E. Information transfer from peptide nucleic acids to RNA by template-directed syntheses. *Nucleic Acids Res.* 1997; 25:4797–4802.

137 SCHMIDT J. G., CHRISTENSEN L., NIELSEN P. E., ORGEL L. E. Information transfer from DNA to peptide nucleic acids by template-directed syntheses. *Nucleic Acids Res.* 1997; 25:4792–796.

138 CHINNERY P. F., TAYLOR R. W., DIEKERT K., LILL R., TURNBULL D. M., LIGHTOWLERS R. N. Peptide nucleic acid delivery to human mitochondria. *Gene Therapy* 1999; 6:1919–1928.

139 RICHARD J. P., MELIKOV K., VIVES E., RAMOS C., VERBEURE B., GAIT M. J., CHEMOMORDIK L. V., LEBLEU B. Cell-penetrating peptides. *J. Biol. Chem.* 2003; 278:585–590.

t1 NIELSEN P. E., EGHOLM M, BERG R. H., BUCHARDT O. *Science* 1991; 254:1497–1500.

t2 HYRUP B., EGHOLM M., NIELSEN P. E., WITTUNG P., NORDÉN B., BUCHARDT O. *J. Am. Chem. Soc.* 1994; 116:7964–7970.

t3 HYRUP B., EGHOLM M., BUCHARDT O., NIELSEN P. E. *Bioorgan. Med. Chem. Lett.* 1996; 6:1083–1088.

t4 KROTZ A.H., LARSEN S., BUCHARDT O., ERIKSSON M., NIELSEN P.E. *Bioorgan. Med. Chem.* 1998; **6**:1983–1992.

t5 KROTZ A.H., BUCHARDT O., NIELSEN P.E. *Tetrahed. Lett.* 1995; **36**:6937–6940.

t6 KROTZ A.H., BUCHARDT O., NIELSEN P.E. *Tetrahed. Lett.* 1995; **36**, 6941–6944.

t7 VAN DER LAAN A.C., VAN AMSTERDAM I., OOSTING R.S., BRANDS R., MEEUWE-NOORD N.J., KUYL-YEHESKIELY E., VAN BOOM J.H. In *Innovation and Perspectives in Solid Phase Synthesis & Combinatorial Synthesis.* 1995, Edinburgh, Scotland, UK, pp. 1–2.

t8 LIOY E., KESSLER H. *Liebigs Ann. Chemie* 1996; 201–204.

t9 KOSYNKINA L., WANG W., LIANG T.C. *Tetrahed. Lett.* 1994; **35**:5173–5176.

t10 HAAIMA G., LOHSE A., BUCHARDT O., NIELSEN P.E. *Ang. Chemie* (International Edition in English) 1996; **35**:1939–1942.

t11 LOWE G., VILAIVAN T. *J. Chem. Soc., Perkin Trans. 1*, 1997; 539–546.

t12 JORDAN S., SCHWEMLER C., KOSCH W., KRETSCHMER A., STROPP U., SCHWENNER E., MIELKE B. *Bioorgan. Med. Chem. Lett.* 1997; **7**:687–690.

t13 HOWARTH N.M., WAKELIN L.P.G. *J. Org. Chem.* 1997; **62**:5441–5450.

t14 SZYRWIEL J., MLYNARZ P., KOZLOWSKI H., TADDEI M. *J. Chem. Soc., Dalton Trans.,* 1998; 1263–1264.

t15 CIAPETTI P., SOCCOLINI F., TADDEI M. *Tetrahedron* 1997; **53**:1167–1176.

t16 FUJII M., YOSHIDA K., HIDAKA J. *Bioorgan. Med. Chem. Lett.* 1997; **7**:637–640.

t17 CANTIN M., SCHÜTZ R., LEUMANN C.J. *Tetrahed. Lett.* 1997; **38**:4211–4214.

t18 SCHÜTZ R., CANTIN M., ROBERTS C., GREINER B., UHLMANN E., LEUMANN C. *Angewandte Chemie* (International Edition in English) 2000; **39**:1250–1253.

t19 ALTMANN K.-H., CHIESI C.S., GARCÍA-ECHEVERRÍA C. *Bioorgan. Med. Chem. Lett.* 1997; **7**:1119–1122.

t20 KUWAHARA M., ARIMITSU M., SISIDO M. *J. Am. Chem. Soc.* 1999; **121**:256–257.

t21 BERGMEIER S.C., FUNDY S.L. *Bioorgan. Med. Chem. Lett.* 1997; **7**:3135–3138.

t22 PEYMAN A., UHLMANN E., WAGNER K., AUGUSTIN S., BREIPOHL G., WILL D.W., SCHÄFER A., WALLMEIER H. *Angewandte Chemie* (International Edition in English) 1996; **35**:2636–2638.

t23 UHLMANN E., WILL D.W., BREIPOHL G., PEYMAN A., LANGNER D., KNOLLE J., O'MALLEY G. *Nucleosides Nucleotides* 1997; **16**:603–608.

t24 EFIMOV V.A., CHOOB M.V., BURYAKOVA A.A., KALINKINA A.L., CHAKHMAKHCHE-VA O.G. *Nucleic Acids Res.* 1998; **26**:566–575.

t25 EFIMOV V.A., CHOOB M.V., BURYAKOVA A.A., CHAKHMAKHCHEVA O.G. *Nucleosides Nucleotides* 1998; **17**:1671–1679.

t26 ZHANG L., MIN J., ZHANG L. *Bioorgan. Med. Chem. Lett.* 1999; **9**:2903–2908.

t27 LIU Y., HUDSON R.H.E. *Synlett* 2001; 1626–1628.

t28 LAGRIFFOUL P., NIELSEN P.E., BUCHARDT D. In 1996, pp. 1–36, Denmark.

t29 JORDAN S., SCHWEMLER C., KOSCH W., KRETSCHMER A., SCHWENNER E., STROPP U., MIELKE B. *Bioorgan. Med. Chem. Lett.* 1997; **7**:681–686.

t30 TSANTRIZOS Y.S., LUNETTA J.F., BOYD M., FADER L.D., WILSON M.-C. *J. Org. Chem.* 1997; **62**:5451–5457.

t31 FADER L.D., BOYD M., TSANTRIZOS Y.S. *J. Org. Chem.* 2001; **66**:3372–3379.

t32 FADER L.D., TSANTRIZOS Y.S. *Org. Lett.* 2002; **4**:63–66.

t33 D'COSTA M., KUMAR V.A., GANESH K.N. *Org. Lett.* 1999; **1**:1513–1516.

t34 D'COSTA M., KUMAR V.A., GANESH K.N. *Org. Lett.* 2001; **3**:1281–1284.

t35 PÜSCHL A., BOESEN T., TEDESCHI T., DAHL O., NIELSEN P.E. *J. Chem. Soc., Perkin Trans. 1*, 2001; 2757–2763.

t36 WADA T., MINAMIMOTO N., INAKI Y., INOUE Y. *J. An. Sci. Technol.* 2002; **122**:6900–6910.

t37 SHIGEYASU M., KUWAHARA M., SHISIDO M., ISHIKAWA T. *Chem. Lett.* 2001; 634–635.

t38 KUMAR V., PALLAN P.S., MEENA M., GANESH K.N. *Org. Lett.* 2001; **3**:1269–1272.

t39 MEENA M., KUMAR V.A., GANESH K.N. *Nucleosides, Nucleotides Nucleic Acids* 2001; **20**:1193–1196.

t40 LI Y., JIN T., LIU K. *Nucleosides, Nucleotides Nucleic Acids* 2001; **20**:1705–1721.

t41 SLAITAS A., YEHESKIELY E. *Nucleosides, Nucleotides Nucleic Acids* 2001; **20**:1377–1379.

t42 LONKAR P. S., KUMAR V. A., GANESH K. N. *Nucleosides, Nucleotides Nucleic Acids* 2001; **20**:1197–1200.

t43 KUMAR V. A., D'COSTA M., GANESH K. N. *Nucleosides, Nucleotides Nucleic Acids* 2001; **20**:1187–1191.

t44 BREGANT S., BURLINA F., VAISSERMANN J., CHASSAING G. *Eur. J. Org. Chem.* 2001; 3285–3294.

t45 VILAIVAN T., SUPARPPROM C., HARNYUTTANAKORN P., LOWE G. *Tetrahed. Lett.* 2001; **42**:5533–5536.

t46 EFIMOV V. A., CHOOB M., BURYAKOVA A. A., PHELAN D., CHAKHMAKHCHEVA O. G. *Nucleosides, Nucleotides Nucleic Acids* 2001; **20**:419–428.

t47 FALKIEWICZ B., WISNIOWSKI W., KOLODZIEJCZYK A. S., WISNIEWSKI K. *Nucleosides, Nucleotides Nucleic Acids* 2001; **20**:1393–1397.

t48 FALKIEWICZ B., KOLODZIEJCZYK A. S., LIBEREK B., WISNIEWSKI K. *Tetrahedron* 2001; **57**:7909–7917.

t49 PÜSCHL A., SFORZA S., HAAIMA G., DAHL O., NIELSEN P. E. *Tetrahed. Lett.* 1998; **39**:4707–4710.

t50 EGHOLM M., CHRISTENSEN L., DUEHOLM K. L., BUCHARDT O., COULL J., NIELSEN P. E. *Nucleic Acids Res.* 1995; **23**:217–222.

t51 HAAIMA G., HANSEN H. F., CHRISTENSEN L., DAHL O., NIELSEN P. E. *Nucleic Acids Res.* 1997; **25**:4639–4643.

t52 ELDRUP A. B., DAHL O., NIELSEN P. E. *J. Am. Chem. Soc.* 1997; **119**:11116–11117.

t53 EGHOLM M., NIELSEN P. E., BUCHARDT O., DUEHOLM K. L., CHRISTENSEN L., COULL J. M., KIELY J., GRIFFITH M. In *World International Property Organization,* 1995, IS. I.S Pharmaceuticals pp. 39–72.

t54 FERRER E., SHEVCHENKO A., ERITJA R. *Bioorgan. Med. Chem.* 2000; **8**:291–297.

t55 CLIVIO P., GUILLAUME D., ADELINE M.-T., HAMON J., RICHE C., FOURREY J.-L. *J. Am. Chem. Soc.* 1998; **120**:1157–1166.

t56 CLIVIO P., GUILLIAUME D., ADELINE, M.-T., FOURREY J.-L. *J. Am. Chem. Soc.* 1997; **119**:5255–5256.

t57 CHALLA H., STYERS M. L., WOSKI S. A. *Org. Lett.* 1999; **1**:1639–1641.

t58 ZHANG B. P., EGHOLM M., PAUL N., PINGLE M., BERGSTROM D. E. *Methods (Orlando, FL)* 2001; **23**:132–140.

t59 LOHSE J., DAHL O., NIELSEN P. E. *Proc. Natl Acad. Sci. USA* 1999; **96**:11804–11808.

t60 HANSEN H. F., CHRISTENSEN L., DAHL O., NIELSEN P. E. *Nucleosides Nucleotides* 1999; **18**:5–9.

t61 CHRISTENSEN L., HANSEN H. F., KOCH T., NIELSEN P. E. *Nucleic Acids Res.* 1998; **26**:2735–2739.

t62 GANGAMANI B. P., KUMAR V. A., GANESH K. N. *Chem. Commun.* 1997; **19**:1913–1914.

t63 ELDRUP A. B., CHRISTENSEN C., HAAIMA G., NIELSEN P. E. *J. Am. Chem. Soc.* 2002; **124**:3254–3262.

t64 ELDRUP A. B., NIELSEN B. B., HAAIMA G., RASMUSSEN H., KASTRUP J. S., CHRISTENSEN C., NIELSEN P. E. *Eur. J. Org. Chem.* 2001, **9**:1781–1790.

t65 IKEDA H., YOSHIDA K., OZEKI M., SAITO I. *Tetrahed. Lett.* 2001; **42**:2529–2531.

t66 OKAMOTO A., TANABE K., SAITO I. *Org. Lett.* 2001; **3**:925–927.

t67 IKEDA H., NAKAMURA Y., SAITO I. *Tetrahed. Lett.* 2002; **43**:5525–5528.

t68 HUDSON R. H. E., LI G., TSE J. *Tetrahed. Lett.* 2002; **43**:1381–1386.

t69 SHIBATA N., DAS B. K., HONJO H., TAKEUCHI Y. *J. Chem. Soc., Perkin Trans. 1* 2001; 1605–1611.

t70 AUSÍN C., ORTEGA J.-A., ROBLES J., GRANDAS A., PEDROSO E. *Org. Lett.* 2002; **4**:4073–4075.

t71 RAJEEV K. G., MAIER M. A., LESNIK E. A., MANOHARAN M. *Org. Lett.* 2002; **4**:4395–4398.

Appendix

Tab. A4.1 Linear PNA analogs

Entry #	Structure	Backbone	ΔT_m DNA °C	ΔT_m RNA °C	Reference
1		Ethylglycine aegPNA	0	0	[t1]
2		Propylglycine	−8.0	v6.5	[t2]
3		Ethyl-β-alanine	−10	−7.5	[t2]
4		Propionyl linker	−20	−16	[t2]
5		Ethyl linker	−22	−18	[t3]
6		Retro inverso	−6.5	n.d.	[t4–t6]
9		L-Ornithine	n.d.	−8	[t7, t8]
10		2-Me-ethylglycine	n.d.	n.d.	[t9]
11		Ethyllysine	L-Lys: −1 D-Lys: +1	L-Lys: −1.3 D-Lys: 0	[t10]

Tab. A4.1   (cont.)

Entry #	Structure	Backbone	ΔT_m DNA °C	ΔT_m RNA °C	Reference
12		Glycine backbone/ ethyl linker	n.d.	n.d.	[t11]
13		Glycylglycine/ ethyl linker	−12.5	n.d.	[t12–t14]
14		Glycine/ ethyl linker	n.d.	n.d.	[t15]
15		β-Amino-alanine	−3.6	n.d.	[t16]
16		E-OPA	−6.5	n.d.	[t17, t18]
17		Z-OPA	−14.2	n.d.	[t17, t18]
18		Serinol-ethyl-methyl linker	n.d.	−2.5	[19]
19		Serinol-ethyl-ethyl linker	n.d.	−3	[t19, t20]
20		α-Methyl-serinol-ethyl-ethyl linker	n.d.	−1.6	[t19]
21		Aminopentan-ethyl linker	n.d.	n.d.	[t21]

Tab. A4.1 (cont.)

Entry #	Structure	Backbone	ΔT_m DNA °C	ΔT_m RNA °C	Reference
22		Hydroxyethyl-phosphono-glycine	Context dependent	Context dependent	[t22–t25]
23		Aminoethyl-phosphono-glycine	Context dependent	Context dependent	[t24, t25]
24		Lysine	n.d.	n.d.	[t26]
25		SNA	n.d.	n.d.	[t27]

Tab. A4.2 Conformationally constrained analogs

Entry #	Structure	Backbone	ΔT_m DNA °C	ΔT_m RNA °C	Reference
26		(S,S)-cyclohexyl	−1.3	0.5	[t28]
27		(R,R)-cyclohexyl	−7	−7.5	[t28]
28		L-Proline	n.d.	n.d.	[t11]
29		D-Proline	n.d.	n.d.	[t11]

Tab. A4.2 (cont.)

Entry #	Structure	Backbone	ΔT_m DNA °C	ΔT_m RNA °C	Reference
30		L-4-trans-amino proline	+2 (towards C-terminus) +2 (towards N-terminus)	n.d.	[t12]
31		L-4-cis-amino proline	−14	n.d.	[t12]
32		D-4-trans-amino proline	−7	n.d.	[t12]
33		β-alanine/proline	+10	n.d.	[t12]
34		Proline-glycine	No hybridization	No hybridization	[t29]
35		APNA	No hybridization	n.d.	[t30]
36		APNA	−12	−14	[t31, t32]

Tab. A4.2 (cont.)

Entry #	Structure	Backbone	ΔT_m DNA °C	ΔT_m RNA °C	Reference
37		APNA	No hybridization	−38	[t31, t32]
38		APNA	−30	−34	[t31, t32]
39		APNA	−26	−25	[t31, t32]
40		Aminoethyl Prolyl (aepPNA)	(2S, 4S): +13 (2R, 4S): +13	n.d.	[t33, t34]
41		(3R, 6R) Piperidone PNA	−19	−10	[t35]
42		(3S, 6R) Piperidone PNA	−12.5	−11.5	[t35]
43		Peptide ribonucleic acids	+1	n.d.	[t36]
44		Cis-L-prolinol PNA	−2	n.d.	[t37]

Tab. A4.2 (cont.)

Entry #	Structure	Backbone	ΔT_m DNA °C	ΔT_m RNA °C	Reference
45		Pyrrolidine PNA	−7	n.d.	[t38]
46		Prolyl carbamate nucleic acids	Central: no N-terminus: −0.5 C-terminus: −13	n.d.	[t39]
47		aepPNA	−2.7	n.d.	[t40]
48		pmgPNA	n.d.	S: −14 R: −8.5	[t41]
49		Pipecolyl PNA	C-terminus: −11 Internal: −21	n.d.	[t42]
50		aepPNA	Antiparallel: +1.5 Parallel: −3.8	n.d.	[t33, t34, t43]
51		amtPNA (Anti)	−17	−15	[t44]
52		amtPNA (Syn)	−22	−15	[t44]

Tab. A4.2 (cont.)

Entry #	Structure	Backbone	ΔT_m DNA °C	ΔT_m RNA °C	Reference
53		Pyrrolidinyl PNA	n.d.	n.d.	[t45]
54		Pyrrolidinyl PNA	n.d.	n.d.	[t45]
55		Pyrrolidinyl PNA	n.d.	n.d.	[t45]
56		Pyrrolidinyl PNA	n.d.	n.d.	[t45]
57		Hydroxy-proline-phosphono PNA (HypNA-pPNA)	n.d.	−1	[t46]

Tab. A4.3 (Amino acid (L-isomer) derived PNA backbones [a)]

Entry [b)]	Backbone modification	T_m (RNA antiparallel)	T_m (DNA antiparallel)
a (1)	Glycine [3]	54.0	50.5
B	Arginine	49.0	45.5
C	Leucine	49.0	45.5
D	Glutamine	49.0	43.5
e (11)	Lysine	48.5	46.0
F	Tyrosine	47.5	42.5
G	Histidine	47.5	42.5
H	Threonine	47.5	44.0
I	Tryptophan	47.0	42.5
J	Phenylalanine	46.5	42.0
K	Valine	46.5	42,0
L	Alanine	49.0	46.5
m (11)	Lysine	47.5	47.0
N	Leucine	47.0	45.0
O	Histidine	46.0	42.0

a) (cf. [t47–t49]).
b) Entry a–k, only one modified monomer was incorporated in the following sequence: H-GTAGAT$_X$CACT-NH$_2$. Entry l–o, three modified monomers were incorporated in the sequence: H-GT$_X$AGAT$_X$CACT$_X$-NH$_2$.
c) aegPNA.

Tab. A4.4 Nucleobase-modified PNA analogs

Entry #	Structure	Nucleobase	ΔT_m DNA °C	ΔT_m RNA °C	Reference
N1		Pseudo iso cytosine	Dependent on Watson–Crick or Hoosteen	n.d.	[t50]
N2		Diamino purine	+3–+5	+2.5–+6	[t51]
N3		3-Oxo-2,3-dihydro-pyridazine	Hoogsteen T-recognition	n.d.	[t52]

Tab. A4.4 (cont.)

Entry #	Structure	Nucleobase	ΔT_m DNA °C	ΔT_m RNA °C	Reference
N4		Isocytosine	n.d.	n.d.	[t53]
N5		5-Bromouracil	n.d.	n.d.	[t53, t54]
N6		5-Methyl cytosine		n.d.	[t54]
N7		4-Thiothymine	n.d.	n.d.	[t55, t56]
N8		$N^{c)}$-Methyl-4-thio-thymin	n.d.	n.d.	[t55]
N9		5-Nitroindole	−9	n.d.	[t57]
N10		3-Nitropyrrole	−13	n.d.	[t57, t58]
N11		2-Thiouracil	−2.5	n.d.	[t59]
N12		Thioguanine	Antiparallel: −8.5 Parallel: −7.5	n.d.	[t60]

Tab. A4.4   (cont.)

Entry #	Structure	Nucleobase	ΔT_m DNA °C	ΔT_m RNA °C	Reference
N13		$N^{d)}$-Benzoyl cytosine	−2.5°C	n.d.	[t61]
N14		2-Amino purine	−2–3	n.d.	[t62]
N15		1,8-Napthyridin-2 (1 *H*)-one	0	+2.5	[t63, t64]
N16		7-Chloro-1,8-napthyridin-2(1*H*)-one	+3	+2.5	[t63, t64]
N17		6-Chloro-1,8-napthyridin-2(1*H*)-one	+2	0	[t63, t64]
N18		7-Methyl-1,8-napthyridin-2(1*H*)-one	+3	+2	[t63, t64]
N19		6-Methyl-1,8-napthyridin-2(1*H*)-one	0	−1	[t63, t64]
N20		5-Methyl-1,8-napthyridin-2(1*H*)-one	+2	0	[t63, t64]

Tab. A4.4 (cont.)

Entry #	Structure	Nucleobase	ΔT_m DNA °C	ΔT_m RNA °C	Reference
N21		Phenothiazine	0	−0.5	[t63, t64]
N22		Flavin	n.d.	n.d.	[t65]
N23		Psoralen	Center: no hybridization N-terminus: +8 C-terminus: +7	n.d.	[t66]
N24		Naphtalimide	Center: −29.3 N-terminus: +0.5 C-terminus: −7.8		[t67]
N25		"PropynylU"	n.d.	n.d.	[t68]
N26		"PropynylU"	n.d.	n.d.	[t68]
N27		"PropynylU"	n.d.	n.d.	[t68]

Tab. A4.4   (cont.)

Entry #	Structure	Nucleobase	ΔT_m DNA °C	ΔT_m RNA °C	Reference
N28		"PropynylU"	n.d.	n.d.	[t68]
N29		Difluoro-phenyl	n.d.	n.d.	[t69]
N30		Perfluoro-phenyl	n.d.	n.d.	[t69]
N31		Difluoro-methyl-phenyl	n.d.	n.d.	[t69]
N32		G-clamp	+10–18		[t70, t71]

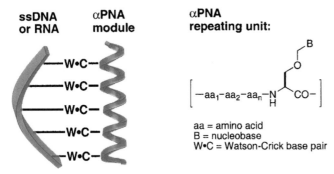

Fig. 5.2 The α-helical peptide nucleic acid (αPNA) design concept. (Reprinted with permission from Garner P., Dey S., Huang Y. *J. Am. Chem. Soc.* **2000**, *122*, 2405 [51])

are attached to the peptide backbone via a flexible methylene link to serine in order to preserve the *N*-glycosidic (O-C-N) substructure found in nucleic acids. In addition to imparting lateral conformational flexibility to the αPNA, the oxygen atom provides an additional H-bond acceptor site that could favorably influence binding and solubility properties. We first considered the tetrad or (i, $i+4$) nucleobase spacing motif. In tetrad αPNAs made up of L-amino acids, the nucleobase containing residues trace a right-handed superhelix that can become aligned by tightening the α-helix to a 3_{10} helix. Assuming that the αPNA has a standard helical pitch of 5.4 Å, synchronized tilting of the nucleobases relative to the α-helix axis would be necessary to approximate the nucleobase spacing of a single-stranded nucleic acid assumed to be in a B-helical geometry (helix rise per base pair = 3.4 Å). In our αPNA structure, the flexible -CH_2-O-CH_2- linker acts as a hinge for this purpose. This, along with the known deformability of both DNA and peptide α-helices in complexes [29], was expected to lead to an induced fit driven by Watson–Crick base-pairing. A molecular modeling study indicated that our αPNA design was geometrically reasonable. That is, a complex formed between αPNA and single-stranded DNA retained its Watson–Crick base pairs after energy minimization.

With the general αPNA design concept formulated, we next turned to deciding exactly what ancillary amino acids should be incorporated into the peptide backbone. We sought a peptide sequence that would be predisposed to α-helix formation to reduce the entropic penalty associated with the peptide coil–helix transition. During the period from 1995 to 1997, Baltzer's group developed a series of peptides designed to fold into a helix–loop–helix motif [30, 31]. The primary amino acid sequence and underlying design rationale of their prototype peptide SA-42 is shown below in Fig. 5.3 A. They found that minor changes in the amino acid sequence did not significantly alter the overall structure of the peptide in solution. For example, KO-42, in which Lys2 and Ala10 of SA-42 had been replaced by histidine residues, still formed a coiled-coil dimer in solution. Our initial goal was to

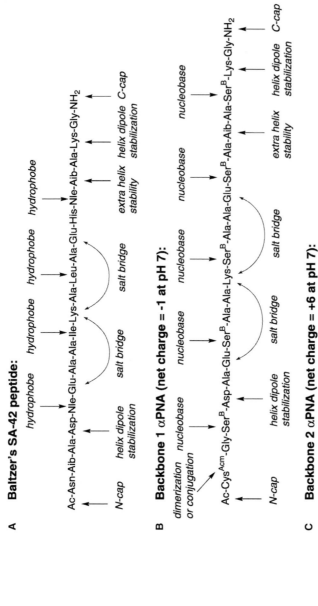

Fig. 5.3 Evolution of the αPNA backbone design. Rationale for choice of amino acid in backbones 1 and 2

minimally edit a known peptide sequence that favors a-helix formation by incorporating the nucleoamino acid residues so that the overall structure of the helix would remain unaffected.

The amino acid sequence of our first aPNA (which we termed backbone 1 or b1) was designed based on this amphipathic helix sequence (Fig. 5.3 B). Specifically, this aPNA backbone included hydrophobic amino acids (Ala and Aib), internal salt bridges (Glu-(aa)$_3$-Lys-(aa)$_3$-Glu), a macrodipole (Asp-(aa)$_{15}$-Lys), and an N-acetyl cap to favor a-helix formation. The C-termini of these aPNA modules end in a carboxamide function to preclude any potential intramolecular end effects. Each aPNA module incorporates five nucleobases for Watson–Crick base pairing to a target nucleic acid sequence.

The incorporation of a Cys residue at position 21 provided the option to link two aPNA modules at their N-terminus (so-called "tail-to-tail" dimer: *CN*-SS-*NC*) via a disulfide bridge using Goddard's "disulphide stitchery" strategy [28]. We decided on a disulfide linkage first to demonstrate validity, even though the resulting aPNA-dimer would probably not be suitable as a drug candidate due to the ease of reductive disulfide cleavage in the cytoplasm. More robust tethers can be accessed via alkylation of Cys (or via other linking strategies). This residue also provides a site for aPNA conjugation to fluorescent probes. Replacement of Gly1 with Cys and Cys21 with Aib provided the complementary b1′ aPNA backbone for disulfide connection at the C-terminus (so-called "head-to-head" dimer: *NC*-SS-*CN*).

While the binding properties of backbone 1 aPNAs with complementary ssDNAs were being studied, it was discovered that the rate of association was extremely slow despite relatively high stability of the resulting aPNA·DNA complexes. This was evident from the lack of a cooperative transition during the cooling cycle of the UV melting experiment. We hypothesized that the negative charge of backbone 1 aPNAs imposed a kinetic barrier to its association with polyanionic oligonucleotide targets. It was surmised that if one could make aPNAs positively charged this kinetic barrier could be reduced by virtue of attractive coulombic forces. In this context, Corey had reported that the attachment of a positively charged lysine-rich peptide to DNA enhanced the rate of duplex formation 48,000-fold [32]. A new backbone was designed based on the simple logic of replacing the negatively charged Asp and Glu residues in backbone 1 with positively charged Lys residues. Additional Lys residues were incorporated at Ala4 and Gly20 and the Aib residue was replaced by Ala. Finally, the C-terminal Gly and Lys were switched in order to avoid close proximity of two Lys residues. The resulting backbone 2 or b2 aPNAs (Fig. 5.3 C) would now have a net charge of +6 at neutral pH.

5.3
Synthesis of αPNAs

5.3.1
Nucleoamino Acid Synthesis

In terms of their molecular structures, the nucleotide and protein realms are usually considered to be rather independent of each other. However, these two families of molecules are covalently linked in the translational aminoacyl-^{t}RNAs and ribonucleoproteins as well as in the nucleoproteins involved in cellular and viral replication. In these hybrid biomolecules, a (deoxy)ribose phosphate moiety serves as the structural connection between the nucleoside and peptide moieties.

Simpler hybrid constructs consisting of nucleoside bases connected directly to amino acid side-chains – namely the nucleoamino acids – are also found in Nature. As early as 1959, Gmelin isolated the nucleoamino acid willardiine, N^1-thyminyl-L-alanine, from *Acacia Willardiana Rose* [33]. A more recent example is the N^9-adenyl-L-α-aminobutyric acid component of FK 374200, a dipeptide metabolite isolated from *Talaromyces* species [34]. Nucleoamino acids based on both natural products containing all four natural DNA nucleobases have been synthesized [35, 36] and incorporated into peptide-based oligonucleotide surrogates. Since the nucleobases were linked to the α-carbon via a simple alkyl chain, the synthesis of these nucleoamino acids could be effected via relatively straightforward pyrimidine and purine alkylation chemistry.

Based on the αPNA design considerations discussed in Section 5.2, the thymine and cytosine containing $N^α$-9-fluorenylmethoxycarbonyl ($N^α$-Fmoc)-protected nucleoamino acids **6** and **9** were targeted and synthesized (Scheme 5.1) by a modification of a route we had used to prepare the analogous $N^α$-t-butoxycarbonyl ($N^α$-Boc)-protected nucleoamino acids [23, 37]. The change in protecting group strategy was necessary in order to accommodate an Fmoc-based solid phase peptide synthesis (SPPS) [38]. Our goal was to synthesize the protected nucleoamino acids in the least number of steps from readily available starting materials.

$N^α$-Fmoc serine benzyl ester **2**, which could be prepared as shown or purchased commercially, was smoothly converted to the crystalline O-methylthiomethyl (MTM) ether **3** in high yield via a Pummerer-like reaction using benzoyl peroxide and dimethyl sulfide in acetonitrile [39]. This common intermediate was used to synthesize both **5** and **8** [40]. Both Ogilvie [41] and Tsantrizos [42] had reported that I_2 was an effective activator with similar MTM ether substrates. The I_2 promoted nucleosidation reaction between O-MTM ether **3** and bis-silylated thymine **4** produced the nucleoamino acid **5** in 60% isolated yield (100% based on recovered **3**). Hydrogenolytic deprotection of the benzyl ester with H_2, Pd/C in MeOH gave the thymine-containing nucleoamino acid **6** in quantitative yield.

To prepare the corresponding cytosine containing nucleoamino acid **9**, N^4-Boc cytosine was first silylated using N,O-bis(trimethylsilyl)acetamide (BSA) under carefully controlled conditions to produce mono-silylated-N^4-Boc cytosine **7**. The mono-silylated product **7** undergoes I_2-mediated nucleosidation with the O-MTM

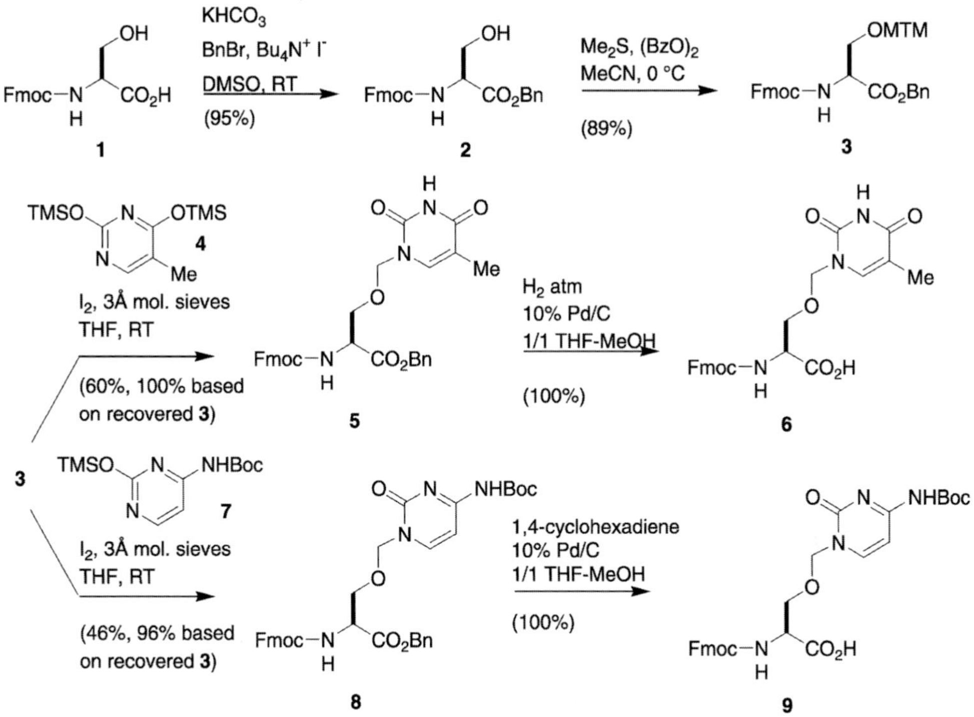

Scheme 5.1 Synthesis of 4-oxonorvalinyl nucleoamino acids

ether **3** to produce a single regioisomer **8** (46% isolated yield, 96% based on recovered **3**). Interestingly, the presence of any bis-silylated cytosine was found to be detrimental to this reaction [43]. Unlike the thymine case, use of Pd/C and H_2 did not result in clean removal of the benzyl group and the formation of side products was observed over time (2–3 h). Clean removal of the benzyl group was observed when 10 equivalents of 1,4-cyclohexadiene was used as a hydrogen transfer reagent along with 10% Pd/C in MeOH/THF (1:1). The cytosine-containing nucleoamino acid **9** was obtained as a pure white solid in 90% isolated yield.

Via these optimized routes, the N-protected pyrimidine nucleoamino acids **6** and **9** were synthesized very efficiently in three high-yielding steps, starting from an inexpensive and commercially available serine derivative. The common intermediate O-MTM ether **3** can be synthesized on a multigram scale and is stable for years if stored in a refrigerator.

Synthesis of the corresponding purine nucleoamino acids Fmoc-Ser$^{A(Boc)}$-OH and Fmoc-Ser$^{G(Boc)}$-OH via the O-MTM ether **3** was compromised by a lack of regioselectivity during the I_2-mediated nucleosidation. The incorporation of these nucleoamino acids into aPNAs is also problematic because of facile acid-catalyzed depurination. The latter problem could be traced to the -CH_2-O-CH_2- linker [38]. We decided to circumvent the problem by synthesizing the isosteric norvalinyl

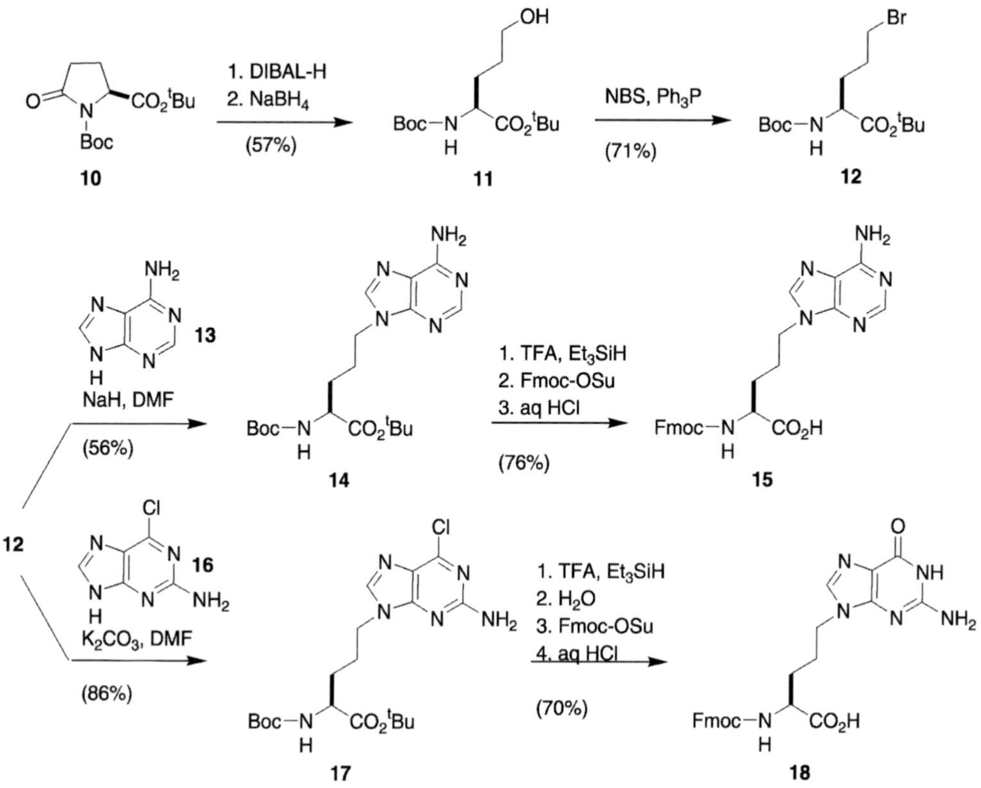

Scheme 5.2 Synthesis of norvalinyl nucleoamino acids

(Nva) nucleoamino acids Fmoc-Nva^A-OH (**15**) and Fmoc-Nva^G-OH (**18**). Both of these nucleoamino acids have now been synthesized regioselectively in good overall yields starting from the known pyroglutamic acid derivative **10** via the sequences shown in Scheme 5.2 [44]. Preliminary experiments have shown that these two nucleoamino acid derivatives can be incorporated into our SPPS without the need for protection of the exocyclic NH$_2$ groups [45].

To summarize, practical syntheses of the protected pyrimidine nucleoamino acids Fmoc-Ser^T-OH (**6**) and Fmoc-Ser$^{C(Boc)}$-OH (**9**) as well as the purine nucleoamino acids Fmoc-Nva^A-OH (**15**) and Fmoc-Nva^G-OH (**18**) have been developed. We are poised to prepare αPNAs containing all four DNA nucleobases.

5.3.2
Solid Phase Synthesis of αPNA

With a satisfactory synthesis of the nucleoamino acids in hand, we next developed a successful strategy for assembling the α-helical peptide nucleic acids on a solid support. Our general synthetic route to the αPNA 21-mer Ac-aa$_{21}$-aa$_{20}$-Ser^B-aa$_{18}$-aa$_{17}$-aa$_{16}$-Ser^B-aa$_{14}$-aa$_{13}$-aa$_{12}$-Ser^B-aa$_{10}$-aa$_9$-aa$_8$-Ser^B-aa$_6$- aa$_5$-aa$_4$-Ser^B-aa$_2$-aa$_1$-NH$_2$ (**21**) and the αPNA-dimers **22**, is depicted in Scheme 5.3 and consisted of three distinct stages: (1) solid phase synthesis of the resin-bound module **20**, (2) peptide cleavage from the resin with concomitant global deprotection of all amino acid residues except cysteine to give the thiol-protected αPNA module **21**, and (3) thiol deprotection and disulfide bond formation to produce the symmetrical dimer **22**. The strategic and tactical issues associated with the solid phase synthesis of αPNAs are very similar to those associated with glycopeptides [46]. Since the CH^α-CH$_2$-O-CH$_2$-B substructure imparts both acid and base sensitivity to our αPNAs [38], we chose the commercially available Rink Amide MBHA resin (Rink amide linker=4-(2,4-dimethoxyphenylaminomethyl) phenoxyacetamido) [47] as the support because αPNA cleavage can be achieved under relatively mild acidic conditions. This resin also exhibits excellent swelling properties and does not lead to by-product formation during cleavage of the peptide from the resin. The I$_2$-labile acetamidomethyl (Acm) protecting group was chosen for the thiol residue of Cys to facilitate a separate disulfide bond formation step [48]. Carpino's HATU (O-(7-azabenzotriazol-1-yl)-1,1,3,3-tetramethyluronium hexafluorophosphate) [49] reagent was used for all peptide couplings (except Fmoc-Cys^{Acm}-OH) and DBU (1,8-diazabicyclo[5.4.0]undec-7-ene) was used for Fmoc deprotection [50]. The amino acid couplings were done in 4:1 NMP (N-methylpyrrolidinone)–DMSO (dimethylsulfoxide) because the solubility of all Fmoc amino acids is superior in NMP compared to DMSO or DMF (N,N-dimethylformamide) and resin swelling was excellent in this co-solvent.

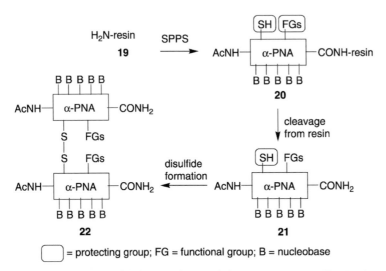

= protecting group; FG = functional group; B = nucleobase

Scheme 5.3 General solid phase synthesis and dimerization strategy. (Reprinted with permission from Garner P, Dey S, Huang Y, Zhang X. *Org. Lett.* **1999**, *1*, 403 [40])

The optimized solid phase synthesis protocol for αPNAs may be summarized as follows. The Fmoc group was cleaved by a 2% solution of DBU in DMF followed by washing with DMF and NMP. HATU-mediated coupling of the amino or nucleoamino acids in the presence of DIEA (*N*,*N*-diisopropylethylamine) was performed in a mixed NMP–DMSO solvent and followed by a DMF wash. An exception was the coupling of Fmoc–Cys^{Acm}-OH, which was effected with DIPCDI (diisopropylcarbodiimide) plus HOBt (hydroxybenzotriazole) to avoid racemization. *N*-terminal acetylation was performed with acetic anhydride and DIEA in DMF followed by washing with DMF and dichloromethane (DCM). Global deprotection of the acid-labile protecting groups and cleavage from the support was accomplished with 95% aqueous TFA. This solid phase protocol was performed in a semi-automated fashion on 100–200 mg of Fmoc-Rink Amide MBHA resin to give 25–50 mg of pure αPNA (about 20% overall yield) after preparative HPLC purification. All αPNAs, as well as the dimers and conjugates described below, were characterized by either MALDI-TOF or ESI mass spectrometry.

Thus, an effective solid phase synthesis of αPNA 21-mers was successfully developed. The resulting SPPS protocols can be applied to all four nucleoamino acids as well as a variety of ancillary amino acids leading to diverse αPNAs structures.

5.3.3
Dimer Formation and Fluorescence Labeling of αPNA

In order to assemble αPNA sequences containing more than five nucleobases, our plan called for joining two αPNA modules together post SPPS via I_2-mediated dimerization of αPNA thiols to obtain the symmetrical αPNA dimers [48]. Exposure of the T_5(b1) monomer to I_2 in MeOH-H_2O provided the tail-to-tail T_5(b1)-dimer. Similar treatment of T_5(b1′) monomer resulted in the head-to-head T_5(b1′)-dimer. However, when we applied these conditions to cytosine-containing backbone 2 αPNAs, a complex reaction profile resulted. It was hypothesized that the oxidation of the free amino groups by I_2 was compromising the desired reaction. This problem was effectively overcome by simply carrying out the reaction in 0.5 N HCl (to convert all of the free amines to their ammonium salts). After their purification by HPLC, the αPNA dimers were characterized by either MALDI-TOF or ESI mass spectrometry. These purified αPNA disulfide dimers also served as starting materials for the synthesis of fluorescence labeled αPNAs.

The synthesis of αPNA derivatives possessing a covalently attached fluorescent tag was necessary for the rapid screening of αPNA libraries as well as the monitoring of cell uptake experiments such as the one to be described in Section 5.5. Post SPPS fluorescence labeling was accomplished using a thiol-reactive dye. We initially chose a dipyrrometheneboron difluoride "BODIPY" fluorescent probe since it has no residual charge that may influence the hybridization of the αPNA and nucleic acid. The procedure for preparing fluorescence-labeled αPNAs involved reducing the disulfide dimer (αPNA-SS-αPNA, prepared as described above) with dithiothreitol (DTT) to give the free thiol (*a*PNA-SH). This is followed by treatment of this thiol with the commercially available BODIPY® FL IA (*N*-(4,4-difluoro-5,7-dimethyl-4-bora-3a,4a-diaza-s-indacene-3-propionyl)-*N'*-iodoacetyle-thylenediamine) reagent.

The orthogonally-protected Cys moiety thus serves as a convenient handle that can be used to link *a*PNA modules together in order to expand its nucleic acid binding capability and/or attach molecular probes.

5.4
*a*PNA-Nucleic Acid Binding Studies

Following the development of an effective *a*PNA synthesis, the binding properties of these novel molecules toward ssDNA were evaluated [40, 51]. DNA was first chosen as the first target to test *a*PNA binding rather than RNA because it can be handled under less stringent conditions and ample binding data exists in the literature. Our primary goal was to validate the *a*PNA hypothesis by demonstrating sequence-specific binding of *a*PNA to its complementary ssDNA. This was accomplished using standard techniques involving thermal denaturation, circular dichroism (CD), and gel mobility retardation.

5.4.1
*a*PNA Secondary Structure

As a prelude to our binding studies, the secondary structure of *a*PNA itself was examined using CD spectroscopy [52]. The first *a*PNA to be studied was the tail-to-tail b1 dimer, [Ac-Cys-Gly-Ser^T-Asp-Ala-Glu-Ser^T-Ala-Ala-Lys-Ser^T-Ala-Ala-Glu-Ser^T-Ala-Aib-Ala-Ser^T-Lys-Gly-NH$_2$]$_2$. The far-UV CD spectra of this *a*PNA in water at 30 °C showed the double minimum at 220 nm (n–π* transition) and 206 nm (π–π* transition) as well as the maximum at 193 nm (π–π* transition), characteristic of a peptide *a*-helix. Upon increasing the temperature, the intensity of the minimum at 200 nm decreased indicating a transition from *a*-helix to random structure. An isodichroic point at 202 nm was suggestive of a temperature-dependent *a*-helix to random coil transition. The helical content of this T$_5$(b1)-dimer at 20 °C in water was estimated to be 26% [40].

Next, the CD spectra of the backbone 2 *a*PNA Ac-Cys^{Acm}-Lys-Ser^C-Ala-Ala-Lys-Ser^C-Ala-Ala-Lys-Ser^T-Ala-Ala-Lys-Ser^C-Ala-Ala-Lys-Ser^C-Gly-Lys-NH$_2$, was measured as a function of pH in phosphate buffer. At pH 7, the secondary structure of this *a*PNA was largely random coil. However, the *a*-helicity of this *a*PNA increased with the pH until it reached a maximum at pH 11. Analogous pH-dependent secondary structure has also been reported for the amphipathic "KALA" peptide Trp-Glu-Ala-Lys-Leu-Ala-[Lys-Ala-Leu-Ala]$_2$-Lys-His-Leu-Ala-Lys-Ala-Leu-Ala-Lys-Ala-Leu$_\varepsilon$-Lys-Ala-Cys-Glu-Ala-OH [53]. In our case however, maximum *a*-helicity occurs at about 50% dissociation (the pK_a of the Lys -NH$_2$ in Ala-Lys oligopeptides is approximately 11.5 in 10 mM NaCl at 0 °C). At pH 12, helicity was lost and the structure began to look more like a β-sheet. We hypothesize that, as the ammonium ions become deprotonated, the charge-repulsion between the neighboring lysine residues decreases and the α-helix secondary structure is favored. The nucleobase could also be influencing the secondary structure of our αPNAs by way of an interaction with the Lys side-chains.

These studies showed that, in the absence of nucleic acid, the backbone 1 αPNA had significant α-helical content at pH 7 whereas the backbone 2 αPNA was largely in a random coil conformation at physiological pH. The latter αPNA did become α-helical at higher pHs in a manner reminiscent of the structurally related amphipathic peptides.

5.4.2
Thermal Denaturation Studies

The UV melting experiment is the traditional method used to evaluate the mutual affinity of nucleic acid surrogates and ssDNA or RNA. This method is based on the well-known hypochromic UV absorption shift of stacked versus unstacked base pairs. Cooperative binding is generally indicated by a sigmoidal absorption versus temperature curve with the transition midpoint being defined as the melting temperature (T_m). Thermodynamic parameters may be derived from the UV melting curve if a two-state binding model is valid. Comparison of the dissociation (heating) and association (cooling) curves also provided a qualitative indication of the binding kinetics. It was for these reasons that we examined our αPNA–DNA complexes first using this technique. Unless otherwise indicated, our UV melting experiments were performed in TE buffer (10 mM Tris-HCl, 1 mM EDTA, pH 7.0) to mimic physiological conditions.

Initially, homobasic DNAs were used for these binding studies. However, DNAs made up of contiguous tracts of guanosine nucleotides such as d(G$_5$) and d(G$_{10}$) tend to form aggregates in solution and are difficult to obtain pure in large quantities. To avoid this problem, dangling nucleobases (generally adenine) were placed on the 5′ and 3′ end of d(G$_5$). These flanking nucleobases are represented by italicized letters. In DNA duplexes, the addition of the dangling bases at both ends increases the T_m of the complex either via additional nonspecific interactions (hydrophobic packing, neighboring base stacking) or the exclusion of solvent intrusion into the terminal base pairs [54]. We found this to be the case with the αPNA·ssDNA complexes as well and subsequently incorporated dangling bases into all of our target oligonucleotide sequences.

An "abasic" peptide (Ac-Cys^{Acm}-Lys-(Ser-Ala-Ala-Lys)$_4$-Ser-Gly-Lys-NH$_2$) with unmodified Ser residues was synthesized and used as a control. No induced cooperative binding to d($TA_3G_5A_3T$) was observed in the UV melting curve with this abasic peptide. This indicated that the cooperative melting between αPNA and its complementary DNA was not merely a reflection of nonspecific interactions be-

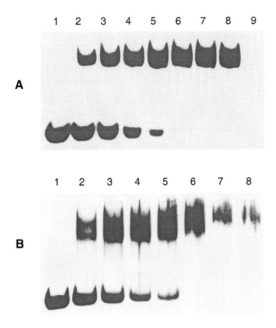

Fig. 5.4 Binding of (**A**) CCTCC(b2) to d(A_3GGAGGA$_3$) and (**B**) T$_5$(b2)-dimer to d($C_3T(TC_2)_2A_{10}C(TC_2)_3$). The ratios of *a*PNA to DNA in (**A**) for lanes 1–8 are 0/1, 1/2, 3/4, 1/1, 5/4, 3/2, 7/4 and 2/1. Lane 9 represents only *a*PNA. The ratios in (**B**) from lanes 1–8 are 0/1, 1/2, 1/1, 3/2, 2/1, 5/2, 3/1, and 4/1. (See [51] for experimental details. Reprinted with permission from Garner P, Dey S, Huang Y. *J. Am. Chem. Soc.* **2000**, *122*, 2405)

sulted in stronger binding. Again, no binding was observed between the "abasic" *a*PNA(b2) peptide Ac-Cys^{Acm}-Lys-(Ser-Ala$_2$-Lys)$_4$-Ser-Gly-Lys-NH$_2$) and d($TA_3G_5A_3T$), underscoring the role that nucleobases play in *a*PNA molecular recognition.

Interestingly, the analogous gel shift experiment with T$_5$(b2)-dimer (net charge +12) and d($C_3T(TC_2)_2A_{10}C(TC_2)_3$) (net charge –29) produced an additional slower-moving species at the expense of the initially-formed complex at *a*PNA/DNA ratios greater than 1 (Fig. 5.4 B). This result is consistent with the formation of a binary 1:1 complex first followed by a ternary 2:1 complex (recall the two-step melting of this complex in Tab. 5.2). It was not possible to distinguish between a tandem duplex or triplex structure for the 2:1 complex.

These gel retardation studies provide unambiguous support for the formation of a 1:1 complex between the b2 *a*PNAs and ssDNA. The dissociation constant (K_d) for these complexes was determined to be in the hundred nanomolar range. In the case of homothymine b2 *a*PNA dimers, an additional 2:1 complex is also evident.

5.4.4
CD Studies on *a*PNA-DNA Complexes

CD spectroscopy has been used extensively to study peptide–nucleic acid interactions. Conformation changes induced by either the peptide or nucleic acid can be detected readily using this spectroscopic technique. Although the CD spectra of peptides and nucleic acids overlap, they do differ significantly in key regions. Peptides have strong CD bands in the 210–230-nm region whereas DNA CD spectra are diagnostic above 240 nm. By focusing on changes in these key regions, one can gain insight into the conformational changes undergone by the component species. The CD studies performed on complexes formed between αPNA and ssDNA not only provided support for the binding event but also for the proposed binding model. Trace b in Fig. 5.5 shows the CD spectrum of αPNA CCTCC(b2) in H_2O in the absence of DNA, which suggests an almost random conformation because of the protonated Lys amines. Upon addition of an equimolar amount of ssDNA d(A_3GGAGGA$_3$), the characteristic CD signatures of an α-helical peptide were observed (Fig. 5.5, trace c: maximum at 191, minima at 206 and 222 nm). Note that, because this is a composite spectrum, the α-helix CD is superimposed upon those of other species present in the sample. We cannot rule out the possibility of co-existing 3_{10}-helix and α-helix structures, since it is not possible to distinguish between them by CD methods [58]. The maximum at 280 nm and minimum at 255 nm are suggestive of an ordered right-handed DNA helix. Since the control CD spectra of αPNA and ssDNA possess different secondary structures, it appears that they are mutually acting as templates for hybridization. Analogous behavior has been noted previously in studies on a peptides corresponding to the basic regions of DNA-binding proteins [59, 60] and the DNA-binding region of the E. coli RecA protein [61]. Thus, a synthetic peptide corresponding to the N-termi-

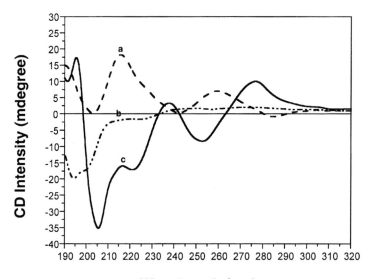

Fig. 5.5 CD spectra of (a) DNA d(A_3GGAGGA$_3$) alone, (b) αPNA CCTCC(b2) alone, and (c) a 1:1 mixture of d(A_3GGAGGA$_3$) + CCTCC(b2) (6 μM each) in H_2O. (See [51] for experimental details. Reprinted with permission from Garner P, Dey S, Huang Y. *J. Am. Chem. Soc.* **2000**, *122*, 2405)

nal 24 amino acid residues of RecA (Rec24) is random in water, while it becomes a-helical in the presence of single-stranded DNA. DNA induces the Rec24 to adopt a highly helical conformation in which the arrangement of positive charge in the DNA-binding peptides is analogous to the 3.6-residue periodicity of hydrophobic amino acid residues in amphipathic a-helices [62].

CD spectroscopy can also be used to determine the stoichiometry of the peptide–nucleic acid interaction [63]. The sample is prepared in a way that allows both the absorption and CD spectra of the oligonucleotide and aPNA to be on the same scale. This is usually done by fixing the total base concentration while varying the ratio of DNA to aPNA. The stoichiometry of the paired strands may be obtained from the mixing curves (Job plots), in which the optical property at a given wavelength is plotted as a function of the mole fraction of each strand. Assuming that the complex and individual components have different CD spectra, an inflection point is observed at the mole percent corresponding to the complex binding stoichiometry.

CD spectra were recorded first for mixtures of aPNA $C_5(b2)$ + DNA $d(TA_3G_5A_3T)$ and $CCTCC(b2)$ + $d(A_3GGAGGA_3)$ at different molar ratios (Fig. 5.6A). A plot of the CD intensity at 258 nm with respect to the mole percentage of aPNA showed two lines of opposite but unequal slope that intersect at a 1:1 molar ratio of the aPNA and DNA strands. We also noted that the complex formed between $T_5(b2)$-

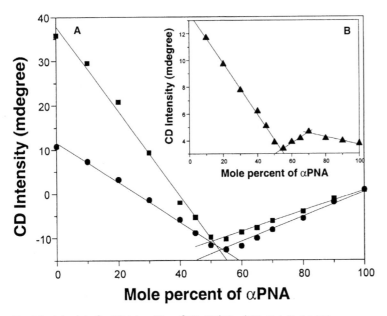

Fig. 5.6 Job plots for CD intensities of **(A)** $C5(b2)$ + $d(TA_3G_5A_3T)$ (at 258 nm, squares), $CCTCC(b2)$ + $d(A_3GGAGGA_3)$ (at 258 nm, circles), and **(B)** $T_5(b2)$-dimer + $d(A_{10})$ (at 261 nm, triangles). (See [51] for experimental details. Reprinted with permission from Garner P, Dey S, Huang Y. *J. Am. Chem. Soc.* **2000**, 122, 2405)

dimer and $d(A_{10})$ showed evidence of both 1:1 and 2:1 binding stoichoimetries (Fig. 5.6 B). This is in line with the UV melting experiment in which two steps in the melting was observed (Tab. 5.2). In contrast, no minimum or maximum was observed for mixtures of total mismatched αPNA $C_5(b2)$ and DNA $d(A_{10})$. This provides the further evidence that αPNA interacts with ssDNA in a sequence-specific manner.

These CD studies confirmed the binding stoichiometries of our αPNA–DNA complexes and provided further support for our binding model. Comparisons between the CD spectra of the individual components and the αPNA·DNA complex suggest a template effect (not unlike that observed with certain DNA-binding proteins) where the components induce mutual conformational changes upon their interaction with each other.

5.5
Pharmacological Studies

Of course, strong and selective binding of a putative drug to its intended target represents an early stage along the drug development pathway. The successful drug candidate must also exhibit satisfactory pharmacological properties. In the context of antisense drug design, both drug delivery and drug stability are important issues that must be addressed. The data presented in this section suggests that the αPNA platform is uniquely suited to the task.

5.5.1
Bioavailability of αPNA

The relatively poor cellular uptake associated with large hydrophilic molecules has posed a major challenge to the development of oligo-based antisense drugs. Over the years, different carrier systems have been developed which improve the penetration rate through lipophilic cell membranes, with liposome-based delivery systems being employed most often. However, recent developments in the area of small peptide-based delivery vehicles, has elicited much excitement in the antisense drug arena [64]. Although peptides and proteins do not generally penetrate cell membranes, a number of polypeptides have been identified that can not only enter cells readily but can even carry covalently attached molecules along with them. These cell-penetrating peptides (CPPs) are generally cationic at physiological pH (due to the presence of lysine and/or arginine residues) and are either derived from specific domains of cell-penetrating and translocating proteins (penetratin, Tat fragment (48–60), signal-sequence-based peptides) or are based on *de novo* peptide designs (transportin, amphipathic model peptides) [65]. A true peptide backbone is not required; totally synthetic constructs based on peptoids [66] and β-peptides [67, 68] can also be used.

In a recent study probing the mechanism by which CPPs gain entry into cells, Hällbrink *et al.* [69] have suggested that CPPs can be divided into two structural classes, one involving amphipathic α-helical peptides with Lys residues contributing the positive charge and the other involving Arg-rich peptides. Members of the first class establish a trans-membrane equilibrium that results in efficient uptake but may also cause membrane leakage, which leads to nonspecific cell death. Interestingly, leakage was found to be proportional to the peptide's hydrophobic moment or amphipathicity. This relationship suggests that a peptide's structure could be modified to maximize delivery and minimize CPP-induced plasma membrane damage and has important implications for αPNA-based drugs.

Since our backbone 2 αPNA incorporates six Lys residues in its peptide sequence and is cationic at a physiological pH, we were optimistic that this αPNA would be taken up into cells without the need for any external carrier system. To answer the simple question of whether b2 αPNAs are internalized, a standard fluorescence microscopy experiment was performed to see if whole cells that were incubated with a fluorescent-labeled αPNA would internalize labeled material [70]. Chinese Hamster Ovary (CHO) cells in culture were incubated with BODIPY-labeled TCCCT(b2) at 37 °C for various periods of time. Following incubation, the cells were rinsed in phosphate-buffered saline (PBS), fixed with 4% formaldehyde at ambient temperature for 20 min, then washed with PBS and stored in a refrigerator until examined by fluorescence microscopy.

Fig. 5.7 shows the fluorescence micrograph of CHO cells incubated with different concentrations of BODIPY-labeled TCCCT(b2) over time at 37 °C. After 30 min, a significant fluorescent signal could be detected in the CHO cells which demonstrated the uptake of αPNAs. Based on a similar study with spermidine, the BODIPY fluorophore does not appear to be sufficient for the internalization of αPNA into CHO cells [71]. From the expanded fluorescence micrograph of CHO cells, it appears that the αPNAs are being distributed more in cytoplasm rather than the nucleus. The experiment also shows that the uptake of αPNAs is

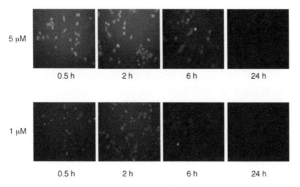

Fig. 5.7 Fluorescence micrograph of CHO cells incubated with different concentrations of BODIPY-labeled αPNA at different times at 37 °C

concentration dependent. At higher concentrations, more aPNA was transported into the cell. It was observed that the level of aPNA accumulation in CHO cells increases over time. However, after 6 h, the fluorescence signal decreases until virtually no aPNA can be detected in the cell after 24 h. This behavior may indicate that the conjugate is being degraded and/or exported during the course of the experiment but it does not affect our conclusions. Importantly, there was no indication of toxicity (cell loss, morphological changes) during these cell uptake experiments.

Oehlke and coworkers have described the cellular uptake properties of a simple α-helical amphipathic model peptide sequence (Lys-Leu-Ala-Leu-Lys-Leu-Ala-Leu-Lys-Ala-Leu-Lys-Ala-Ala-Leu-Lys-Leu-Ala) in the context of a drug delivery vehicle [72]. On the basis of the data presented, it was proposed that non-endocytosis mechanism(s) were involved in the uptake into mammalian cells. The similarity between our b2 aPNA-sequence to that of this amphipathic model peptide makes it tempting to suggest that a similar uptake mechanism is involved in the cellular uptake of aPNAs. Further experimentation is necessary to test this hypothesis.

Our initial bioavailability experiments indicate that cationic aPNAs are readily internalized by cells without any additional delivery vehicle and that they do not affect cell viability. We are currently examining aPNA internalization with other cell types as well as the intracellular localization of aPNAs.

5.5.2
Biostability of aPNA

While replacement of the ribose phosphate backbone with a peptide backbone will certainly result in stabilizing the oligonucleotide surrogate to nucleases, the peptide backbone may itself be susceptible to enzyme-mediated proteolysis. Since we propose to develop a new approach to antisense therapeutics based on aPNAs, it is important to show that these molecules will not be degraded by proteases prior to reaching their nucleic acid targets. Such degradation would, of course, undermine the therapeutic potential of our aPNAs. To address this concern, we have examined the susceptibility of aPNA to protease degradation by comparing the lifetime of aPNA with that of a control peptide using human serum as a protease source [73]. We chose to conduct the experiment in this manner since measurement of *in vitro* peptide stability towards human serum is a convenient way to estimate peptide stability in the blood [74]. Our precedent for this approach comes from the work of Powell and co-workers, who have used human serum as a protease source to access the *in vitro* stability of small peptide major histocompatibility complex (MHC) antagonists [75]. Demidov *et al.* performed a similar study on Nielsen PNAs and concluded that they were relatively stable towards human serum [76].

Two aPNAs were used in this study, L-CTCCT(b2) as well as its antipode D-CTCCT(b2) made up of unnaturally configured amino acids. A control peptide lacking nucleobases, Ac-Trp-Cys^{Acm}-Lys-Ser-(Ala$_2$-Lys-Ser)$_4$-Gly-Lys-NH$_2$, was also synthesized with the expectation that it would be degraded and thus validate the experiment. In this control peptide, a Trp residue was inserted at the N-terminus to facilitate UV detection in the absence of nucleobase chromophores during HPLC analysis of the degradation. Upon exposure to freshly prepared solutions of human serum, changes in the relative concentrations of the aPNAs as well as the control peptide were followed by analyzing their HPLC profiles over time. The concentration of the control peptide decreased exponentially reaching a baseline value after about 12 min. This experiment demonstrated that our human serum had not been deactivated during its processing and contained active protease. Perhaps not unexpectedly, the unnaturally configured D-CTCCT (made up of all D-amino acids) showed insignificant degradation after exposure to human serum for 6 h. Most surprising, however, was the fact that no degradation was observed when the "naturally-configured" L-CTCCT was exposed to active human serum over this same period of time. In these experiments, we did not detect the formation of any aPNA-derived products.

While the N-acetyl group likely protects the N-terminus from the action of aminopeptidases and the carboxamide moiety protects the C-terminus from carboxypeptidases, these features alone do not prevent degradation of the control peptide. The stability of aPNAs towards proteases is independent of the absolute configuration of the backbone amino acids. The aPNAs used in this study are disordered in the absence of nucleic acid complements; so enforced helicity cannot be a factor in their enhanced metabolic stability. Thus it appears that the unnatural nucleoamino acid residues are primarily responsible for the enhanced stability of L-configured aPNAs towards degradation by proteases in human serum. While it is tempting to ascribe this behavior to the inability of the enzyme to accommodate the nucleobase-containing amino acid side chain, other possible explanations cannot yet be rigorously excluded (enzyme inactivation by aPNA or the formation of stable aPNA–protein complexes, for example).

The preliminary data suggests that aPNAs are stable towards degradation by the proteases present in human serum. Further studies are required to determine the stability of aPNAs *in vivo*.

5.6
Conclusions and Prospects

Although our research program is still at a relatively early stage of development, we believe that the results obtained thus far augur well for the future of antisense drugs based on the aPNA concept. So far, the following milestones have been reached. (1) An effective solid phase synthesis of aPNAs has been achieved that is amenable to the assembly of aPNA libraries for SAR studies. (2) It has been demonstrated that aPNAs bind to their complementary ssDNAs with high affinity and sequence specificity. The aPNA repertoire of binding interactions includes electrostatic and hydrophobic components in addition to the Watson–Crick base pairs.

(3) Cationic aPNAs are taken up into living cells in a dose-dependant manner without affecting cell viability. (4) We have shown that aPNAs are stable to the nucleases present in human serum for up to 6 h.

Much work remains to be done. Although our binding and CD data are consistent with the a-helix-based binding model, we plan to determine the complete 3-D structure of a aPNA·DNA complex using NMR techniques. The optimization of our aPNA structure for binding to RNA is another important goal. Preliminary studies show that aPNA does bind to RNA but with lower affinity than it binds to DNA. One approach to this problem calls for tailoring the electrostatic binding interactions between cationic aPNAs and RNA using combinatorial techniques. The ultimate goal for this phase of the project will be to demonstrate the dose-dependent downregulation of a specifically targeted protein by the administration of an antisense aPNA. The high affinity of aPNAs for their complementary nucleic acids could make a "steric blockade" antisense mechanism viable, which could enhance specificity. If they do not have a significant affinity for proteins, aPNAs could form the basis for a new generation of antisense agents.

The major attribute of aPNA versus other nucleic acid surrogates may well be the ability to incorporate additional chemical functionality via the ancillary amino acids without adversely affecting the primary mode of molecular recognition. It is this potential for structure modification that sets aPNA apart from other antisense drug platforms. Pharmacologically-oriented SAR studies can be envisioned to fine-tune the uptake of cationic aPNAs into specific cell types while minimizing undesired effects. A particularly fascinating possibility is the development of an aPNA that is also a cell-selective delivery vehicle. Precedent for this idea comes from the work of Mor and coworkers, who have shown that antimicrobial dermaseptin peptides can be loaded onto red blood cells (RBCs) and transferred to bacteria via an affinity-driven process [77].

5.7
Addendum

Since our original aPNA publication, there have been other reports of the incorporation of nucleobases into a-helical peptides. Mihara and coworkers reported that a-helical coiled-coiled peptides could be stabilized by base pairs between complementary γ-nucleobase-a-aminobutanoic acids [78] They have also reported that the incorporation of such nucleoamino acids into a-helical segments of HIV-1 Rev and HIV-1 nucleocapsid protein can result in increased binding affinity and specificity to HIV-1 RRE RNA and SL3 RNA respectively [79, 80].

5.8
Acknowledgements

PG would like to thank Dr Yvan Guindon for the opportunity to spend a sabbatical at L'Institute de Recherches Cliniques de Montréal (IRCM). The authors offer special thanks to Drs Guindon and Youssef Bennani (Athersys) for reading over the manuscript and providing constructive feedback as well as their unique insights on drug development. We thank Professor Ruth Siegel (CWRU-Pharmacology) for her assistance with the cell uptake experiments.

5.9
References

1 THOMPSON J. D., *Drug Discov. Today* **2002**, 7, 912–917.
2 ZAMECNIK P. C, STEPHENSEN M. L., *Proc. Natl Acad. Sci. USA* **1978**, 75, 280–284.
3 Antisense technology Part A: General methods, methods of delivery and RNA studies. In PHILLIPS M. I. (Ed.), *Methods in Enzymology* Vol. 313. Academic Press: San Diego, CA, **2000**.
4 Antisense technology Part B: Applications. In PHILLIPS M. I. (Ed.), *Methods Enzymology* Vol. 314. Academic Press: San Diego, CA, **1999**.
5 FIRE A. *et al. Nature* **1998**, 391, 806–811.
6 HUTVAGNER G., Zamore P. D., *Science* **2002**, 297, 2056–2060.
7 ELBASHIR S. M., *et al. Nature* **2001**, 411, 494–498.
8 ANDERSON W. F., *Hum. Gene Ther.* **2002**, 13, 1261–1262.
9 LEBEDEVA I., STEIN C. A., *Annu. Rev. Pharmacol. Toxicol.* **2001**, 41, 403–419.
10 *Antisense Drug Technology – Principles, Strategies and Applications*, CROOKE S. T. (Ed.). M. Decker, **2001**.
11 KING D. J., VENTURA D. A., BRASIER A. R., GORENSTEIN D. G., *Biochemistry* **1998**, 37, 16489–16493.
12 SUMMERTON J., WELLER D., *Antisense Nucleic Acid Drug Dev.* **1997**, 7, 187–195.
13 SUMMERTON J., STEIN D., HUANG S., MATTHEWS P., WELLER D., PARTRIDGE M., *Antisense Nucleic Acid Drug Dev.* **1997**, 7, 63–70.
14 DE KONING H., PANDIT U. K., *Rec. Trav. Chim.* **1971**, 91, 1069–1080.
15 BUTTREY J. D., JONES A. S., WALKER R. T., *Tetrahedron* **1975**, 31, 73–75.
16 SHVACHKIN YU. P., *Zh. Obshch. Khim.* **1979**, 49, 1157–1161.
17 FALKIEWICZ B., *Acta Biochim. Pol.* **1999**, 43, 509–529.
18 NIELSEN P. E, EGHOLM M., BERG R. H., BUCHARDT O., *Science* **1991**, 254, 1497–1500.
19 SOOMETS U., HÄLLBRINK M., LANGEL Ü., *Front. Biosci.* **1999**, 4, d782–d786.
20 SFORZA S., HAAIMA G., MARCHELLI R., NIELSEN P. E., *Eur. J. Org. Chem.* **1999**, 197–204.
21 SFORZA S., CORRADINI R., GHIRARDI S., DOSSENA A., MARCHELLI R., *Eur. J. Org. Chem.* **2000**, 2905–2913.
22 UHLMANN E., PEYMAN A., BREIPOHL G., WILL D. W., *Angew. Chem. Int. Ed.* **1998**, 37, 2796–2823.
23 GARNER P., YOO J. U., *Tetrahed. Lett.* **1993**, 34, 1275–1278.
24 HARRISON S. C., *Science* **1991**, 353, 715–719.
25 BATTISTE J. L., MAO H., RAO N. S., TAN R., MUHANDIRAM D. R., KAY L. E., FRANKEL A. D., WILLIAMSON J. R., *Science* **1996**, 273, 1547–1551.
26 BLUNDELL T., BARLOW D., BORKAKOTI N., THORNTON J., *Nature* **1983**, 306, 281–283.
27 CHAKRABARTI P., BERNARD M., REES D. C., *Biopolymers* **1986**, 25, 1087–1093.

28 PARK C., CAMPBELL J. L., GODDARD W. A., III *J. Am. Chem. Soc.* **1995**, *117*, 6287–6291.

29 CURRAN T., KERPPOLA T. K., *Science* **1991**, *254*, 1210–1214.

30 OLOFSSON S., JOHANSSON G., BALTZER L., *J. Chem. Soc. Perkin Trans. 2* **1995**, 2047–2056.

31 BROO K. S., BRIVE L., AHLBERG P., BALTZER L., *J. Am. Chem. Soc.* **1997**, *119*, 11362–11372.

32 COREY D. R., *J. Am. Chem. Soc.* **1995**, *117*, 9373–9374.

33 GMELIN R., *Hoppe Seyler's Z. Physiol. Chem.* **1959**, *316*, 164–169.

34 MORINO T., NISHIMOTO M., MASUDA A., FUJITA S., NISHIKIORI T., SAITO S., *J. Antibiot.* **1995**, *48*, 1509–1510.

35 LOHSE P., OBERHAUSER B., OBERHAUSER-HOFBAUER A., ESCHENMOSER A., *Croatica Chim. Acta* **1996**, 535–562

36 LENZI A., REGINATO G., TADDEI M., *Tetrahed. Lett.* **1995**, *36*, 1713–1716.

37 YOO J. U., Part I: Studies Directed Towards the Synthesis of Amipurimycin. Part II: The Development of Novel Peptide Based Nucleic Acid Surrogates. PhD thesis, Case Western Reserve University, Cleveland, OH, January **1994**.

38 DEY S., Alpha Helical Peptide Nucleic Acids (Alpha PNAs) – Integration of Protein Structure and Nucleic Acid Function. PhD thesis, Case Western Reserve University, Cleveland, OH, January **2001**.

39 MEDIAN J. C., SALOMON M., KYLER K. S., *Tetrahed. Lett.* **1988**, *29*, 3773–3776.

40 GARNER P., DEY S., HUANG Y., ZHANG X., *Org. Lett.* **1999**, *1*, 403–405.

41 OGILVIE K. K., NGUYEN-BA N., HAMILTON R. G., *Can. J. Chem.* **1984**, *62*, 1622–1627.

42 TSANTRIZOS Y. S., LUNETTA J. F., BOYD M., FADER L. D., WILSON M.-C., *J. Org. Chem.* **1997**, *62*, 5451–5457.

43 HUANG Y., DEY S., GARNER P., *Tetrahed. Lett.* **2003**, *44*, 1441–1444.

44 MICHELLE ADAMS, unpublished data.

45 SETH ENNIS, unpublished data.

46 PAULSEN H., SCHLEYER A., MATHIEUX N., MELDAL M., BOCK K., *J. Chem. Soc. Perkin Trans. 1* **1997**, 281–293.

47 RINK H., *Tetrahed. Lett.* **1987**, *28*, 3787–3790.

48 VEBER D. F., MILKOWSKI J. D., VARGA S. L., DENKEWALTER R. G., HIRSCHMANN R., *J. Am. Chem. Soc.* **1972**, *94*, 5456–5461.

49 CARPINO L. A., *J. Am. Chem. Soc.* **1993**, *115*, 4397–4398.

50 WADE J. D., BEDFORD J., SHEPPARD R. C., TREGEAR G. W., *Peptide Res.* **1991**, *4*, 194–199.

51 GARNER P., DEY S., HUANG Y,. *J. Am. Chem. Soc.* **2000**, *122*, 2405–2406.

52 FASMAN G. D. (Ed.) *Circular Dichroism and the Conformational Analysis of Biomolecules*, Plenum Press: New York, **1996**.

53 WYMAN T. B., NICOL F., ZELPHATI O., SCARIA P. V., PLANK C., SZOKA F. C., JR., *Biochemistry* **1997**, *36*, 3008–3017.

54 FREIER S. M., BURGER B. J., ALKEMA D., NEILSON T., TURNER D. H., *Biochemistry* **1983**, *22*, 6198–6206.

55 GARNER P., HUANG Y., DEY S., *ChemBioChem* **2001**, *2*, 224–226.

56 FERSHT A. *Structure and Mechanism in Protein Science*, Freeman: New York, **1999**.

57 TAYLOR J. D., ACKROYD A. J., HALFORD S. E., The Gel Shift Assay for the analysis of DNA–protein interactions. In *DNA–Protein Interactions: Principles and Protocols. Methods in Molecular Biology* Vol. 30, KNEALE G. G. (Ed.), Humana Press Inc.: Totowa, NJ, **1994**.

58 SUDHA T. S., VIJAYAKUMAR E. K. S., BALARAM P., *Int. J. Pept. Prot. Res.* **1983**, *22*, 464–468.

59 PATEL L., ABATE C., CURRAN T., *Nature* **1990**, *347*, 572–575.

60 WEISS M. A., ELLENBERGER T., WOBBE C. R., LEE J. P., HARRISON S. C., STRUHL K., *Nature* **1990**, *347*, 575–578.

61 ZLOTNICK A., BRENNER S. L., *J. Mol. Biol.* **1988**, *209*, 447–457.

62 DEGRADO W. F., LEAR J. D., *J. Am. Chem. Soc.* **1985**, *107*, 7684–7689.

63 GRAY D. M., HUNG S., JOHNSON K. H., *Method. Enzymol.* **1995**, *246*, 19–34.

64 LEBEDEVA I., BENIMETSKAYA L., STEIN C. A, VILENCHIK M., *Eur. J. Pharm. Biopharm.* **2000**, *50*, 101–119.

65 LINDGREN M., HÄLLBRINK M., PROCHIANTZ A., LANGEL Ü., *Trends Pharmacol. Sci.* **2000**, *21*, 99–103.

66 Wender P.A., Mitchell D.J., Patta-biraman K., Pellkey E.T., Steinman L., Rothbard J.B., *Proc. Natl Acad. Sci. USA* **2000**, *97*, 13003–13008.

67 Umezawa N., Gelman M.A., Haigis M.C., Raines R.T., Gellman S.H., *J. Am. Chem. Soc.* **2002**, *124*, 368–369.

68 Rueping M., Mahajan Y., Sauer M., Seebach D., *ChemBioChem* **2002**, *3*, 257–259.

69 Hällbrink M., Florén A., Elmquist A., Pooga M., Bartfai T., Langel Ü., *Biochim. Biophys. Acta* **2001**, *1515*, 101–109.

70 Huang Y., Development of Antisense Alpha-Helical Peptide Nucleic Acids. PhD thesis, Case Western Reserve Unvierisity, Cleveland, OH, May **2002**.

71 Soulet D., Covassin L., Kaouass M., Charest-Gaudreault R., Audette M., Poulin R., *Biochem. J.* **2002**, *367*, 347–357.

72 Oehlke J., Scheller A., Wiesner B., Krause E., Beyermann M., Lauschenz E., Melzig M., Bienert M., *Biochim. Biophys. Acta* **1998**, *1414*, 127–139

73 Garner P., Sherry B., Moilanen S., Huang Y., *Bioorg. Med. Chem. Lett.* **2001**, *11*, 2315–2317.

74 Powell M.F., *Annu. Rep. Med. Chem.* **1993**, *28*, 285–294.

75 Powell M.F., Grey H., Gaeta F., Sette A., Colón S., *J. Pharm. Sci.* **1992**, *81*, 731–735.

76 Demidov V.V., Potaman V.N., Frank-Kamenetskii M.D., Egholm M., Buchard O., Sönnichsen S.H., Nielsen P.E., *Biochem. Pharmacol.* **1994**, *48*, 1310–1313.

77 Mor A., *Drug Develop. Res.* **2000**, *50*, 440-447.

78 Matsumoto S., Ueno A., Akihiko U., Mihara H., *Chem. Commun.* **2000**, 1615–1616.

79 Takahashi T., Hamasaki K., Ueno A., Mihara H., *Bioorg. Med. Chem.* **2001**, *9*, 991–1000.

80 Takahashi T., Ueno A., Mihara H., *ChemBioChem* **2002**, *3*, 543–549.

6

DNA and RNA-cleaving Pseudo-peptides

Alessandro Scarso and Paolo Scrimin

6.1
Introduction

This chapter focuses on the cleavage of the phosphate bond by catalysts in which peptide sequences, which may be very short, or pseudo-peptides play a relevant role in the process, either by taking part directly in the catalysis or by providing structural features to facilitate the cleavage. In a few cases the cleavage is based on amino acids alone and these will also be discussed. It is intended to provide an overview of the intense scientific activity centered on the role of pseudo-peptides as potential catalysts in a number of reactions including the cleavage of phosphates, rather than a complete and detailed discussion. One of the reasons for this is obviously the fact that among the most powerful catalysts known are enzymes and these are proteins, i.e. polypeptides. In contrast to other bonds the phosphate bond is a fairly tough substrate which makes any endeavor to cleave it even more challenging. Catalysts for the selective cleavage of nucleic acids are extremely interesting for applications ranging from medicinal chemistry (cancer therapy for instance) to gene manipulation.

6.2
Cleavage of a Phosphate Bond

The cleavage of nucleic acid may be accomplished mainly by two pathways: by oxidation (in this case the process is centered on the sugar moiety) or by hydrolysis (in this case the phosphate group is the primary target). The phosphate bond of phosphoric acid diesters is particularly resistant to hydrolytic cleavage under physiological conditions (i.e. at pH close to 7) when these substrates are present as anions [1]. Apart from the intrinsic reactivity of the functional group, the problem is compounded under these conditions by the electrostatic repulsion of any anionic nucleophile (OH^- for instance), a situation that does not apply to the peptide bond. This sluggish reactivity is highlighted by the half-lives reported for the hydrolysis of the most common phosphate bonds present in the biological world,

Pseudo-peptides in Drug Discovery. Edited by Peter E. Nielsen
Copyright © 2004 Wiley-VCH Verlag GmbH & Co. KGaA, Weinheim
ISBN: 3-527-30633-1

those of DNA and RNA [2]. In the former the half-life is $\sim 10^{10}$ years whilst in the latter it is $\sim 10^2$ years. The reason for a phosphate in RNA being more reactive than that in DNA is because bond cleavage is due to an intramolecular attack of the proximal -OH of the ribose in the case of RNA. This situation is obviously not possible with DNA.

Many enzymes responsible for the hydrolysis of RNA or DNA present in their catalytic site a metal ion as cofactor which is essential in the case of hydrolytic DNA cleavage. The presence of ammonium groups provides an important contribution to the catalysis, particularly in the case of RNA [3]. In spite of the fact that the structure of many of these proteins is known from their X-ray crystal structure [4], the design and preparation of catalysts equaling, not to say rivaling, their activity is still a challenging task. Many contributions have unraveled important aspects that may be useful for obtaining good catalysts. With reference to the role played by metal ions this may be summarized as follows [1]. (a) The coordination of the $P(=O)-O^-$ group to the metal ion dissipates the negative charge thus removing the electrostatic repulsion for the attack by OH^- (or any other anionic nucleophile) and provides Lewis acid-like catalysis; (b) coordination of a water molecule decreases its pKa thus providing increased amount of (metal-bound) OH^- at physiological pH; (c) coordination of the leaving group facilitates its departure also by decreasing its pKa. The seminal work by Chin [1] and his associates has provided compelling evidence that by adding together all these contributions, up to a 10^{18}-fold increase in the rate of hydrolysis can be achieved, comparable to that observed with phosphatases. For this purpose at least two metal ions must be present in the catalytic site.

An important aspect of any synthetic catalyst is its ability to place the functional groups relevant to the catalytic process in the correct place. Thus, in the case of two metal ions, their relative distance is of critical importance as well as that of any nucleophile present at the catalytic site. X-ray diffraction studies indicate that for dinuclear phosphatases for instance, the distance between the two metal centers is in the range 3.5–5 Å [4]. This aspect of the demanding challenge of designing a catalyst requires the selection of molecular architectures whose conformation has minimal flexibility but which can nevertheless be easily manipulated. It is not surprising that an increasing number of examples of catalysts designed for the hydrolytic cleavage of phosphate esters is based on peptides: the control of their conformation is becoming accessible and unnatural amino acids [5–7] can be introduced to add to the standard natural pool to increase the number of functional groups available for the construction of the putative catalytic site.

This chapter will focus on such catalysts with particular emphasis to those acting on the basis of a hydrolytic mechanism. This is indeed the most challenging approach although many different mechanisms are known for cleaving DNA and RNA, for instance oxidative systems. With reference to any biological target a hydrolytic mechanism also has the advantage of producing non-toxic fragments and, hence, it may be appealing for such applications as gene manipulation or for the production of antibiotics or other pharmacologically active compounds.

6.3
Systems which Utilize an Oxidative Mechanism

Although, as stated above, we will mostly focus on hydrolytic systems it is worth discussing oxidation catalysts briefly [8]. Probably the best known of these systems is exemplified by the antitumor antibiotics belonging to the family of bleomycins (Fig. 6.1) [9]. These molecules may be included in the list of peptide-based catalysts because of the presence of a small peptide which is involved both in the coordination to the metal ion (essential co-factor for the catalyst) and as a tether for a bisthiazole moiety that ensures interaction with DNA. It has recently been reported that bleomycins will also cleave RNA [10]. With these antibiotics DNA cleavage is known to be selective, preferentially occurring at 5'-GpC-3' and 5'-GpT-3' sequences, and results from metal-dependent oxidation [11]. Thus it is not a cleavage that occurs at the level of a P-O bond as expected for a non-hydrolytic mechanism.

In the case of bleomycins the peptide is only to a minor extent involved in the coordination to the metal centers. Simple peptides specifically designed for this purpose have been studied by Long [12,13] and have the general formula NH_2-Xaa-Xaa-His-$CONH_2$ (Fig. 6.2). For instance the tripeptide Gly-Gly-His represents the consensus sequence of the amino-terminal Cu(II) or Ni(II) chelating domain of the serum albumins. The histidine imidazole nitrogen of this sequence coordinates to the metal centers and, subsequently two intervening deprotonated amide nitrogens provide further binding units. In particular Ni(II)-Xaa-Xaa-His metallopeptides have proven to be unique agents able to interact either with the minor groove of the B-form of DNA or loop regions of structured RNAs. These Ni(II) complexes cleave DNA by abstracting the C4 hydrogen of a deoxyribose. This mechanism parallels that of bleomycins mentioned above. It is proposed that the active species is a peptide-bound Ni(III)-OH or Ni(IV)=O ($KHSO_5$ or H_2O_2 are the oxidizing species present in the solution). The cleavage selectivity is a result of the specificity in binding rather than of any preference for a nucleobase. In the case of RNA, cleavage was primarily observed in loop structures. The metallopeptides appear to interact within a particular loop like many RNA-binding proteins. A similar approach was also pursued by Burrows [14] who studied similar peptide sequences extending the coordination motif with a salen unit. This system, as a Ni(II) complex, combines the chemical reactivity of a salen complex with the potential molecular recognition properties of a peptide.

More sophisticated systems were designed in order to deliver the DNA-cleaving unit to the target with great selectivity. Modification of DNA binding proteins is obviously a rather attractive approach. One of the first examples was reported by Dervan [15] who connected the Gly-Gly-His tripeptide to the amino terminus of the DNA-binding domain of Hin recombinase. The two connected elements have specific roles: the first, as a Cu(II) complex, in the presence of H_2O_2 and sodium ascorbate cleaves the biopolymer while the second provides sequence-specific recognition. Bruice and Sigman and colleagues [16] have converted λ phage Cro protein into a selective nucleolytic agent by linking it to the oxidative chemical nucle-

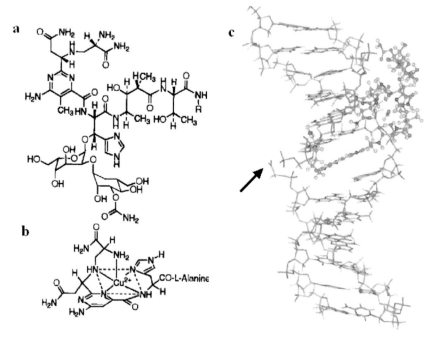

Fig. 6.1 (a) The peptide and ligand subunits of bleomycin, BLM (R denotes the DNA-interacting unit); (b) proposed structure of the complex with Cu^{2+} and part of the peptide involved in binding; (c) average minimized structure of the Co(III)-BLM-OOH bound to phosphoglycolate lesion (arrow) containing oligonucleotide. The BLM is shown in ball-and-stick form. (Reprinted from [11a] with permission from the American Chemical Society)

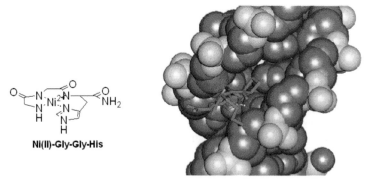

Ni(II)-Gly-Gly-His

Fig. 6.2 The Ni(II)-Gly-Gly-His complex and the molecular model of Ni(II)-Pro-Lys-His bound to the minor groove of an A-T-rich region of B-form DNA. (Reprinted from [12] with permission from the American Chemical Society)

ase 1,10-phenanthroline-copper. An important feature of this semi-synthetic nuclease is that Cro retains its high affinity for the major groove while directing the nucleolytic activity of 1,10 penanthroline-copper towards its chemically-reactive site in the minor groove. Indeed scission was observed within the recognition sequence. In contrast, Sugiura et al. [17] used Cys_2-His_2-type zinc finger proteins and connected to this motif the tripeptide Gly-Gly-His to be used for oxidative DNA cleavage as a Ni(II) complex (see above).

A fully synthetic system was recently described by Barton, Stemp and associates [18]. They have tethered short peptides to a DNA-intercalating ruthenium complex to create a photoactivated cross-linking agent. The ruthenium complex delivers the peptide to DNA [19] and initiates a cross-linking reaction by oxidizing DNA upon irradiation in the presence of an oxidative quencher. The largest degree of cross-linking was observed with peptides containing positively-charged residues, either lysine or arginine (the Lys-X-Lys motif being the best). The cross-linking is achieved in this case through the flash–quench chemistry by using $[Co(NH_3)_5Cl]^{2+}$ as the quencher.

6.4
Systems Based on a Hydrolytic Mechanism

6.4.1
DNA, Oligo- and Polynucleotides, and Synthetic Models as Substrates

Several of the principles that have been applied in the examples illustrated in the previous section have also been used in the preparation of hydrolytically-active catalysts. Thus the amplification of the scope of existing proteins may be achieved via their covalent modification [20]. One approach for targeting DNA or RNA is to conjugate an enzyme to an oligonucleotide, for instance. Schultz and Pei [21] described one such example back in 1991. Previously [22], by using a combination of both chemical and genetic approaches they had been able to convert a relatively non-specific phosphodiesterase, staphylococcal nuclease, into a molecule capable of hydrolyzing RNA, single-stranded DNA and duplex DNA in a sequence-specific manner. In this particular case two semi-synthetic nucleases were synthesized by coupling 13-nucleotide and 12-nucleotide to a K84C staphylococcal nuclease mutant. Because of their constituents the new molecules were able to bind to specific sequences of plasmid pUC19 via triple helix formation and hydrolytically cleave the plasmid into two fragments with very high efficiency. A similar approach was followed by Thuong and co-workers [23].

Slightly more complex systems have been described which involve the conjugation of two different protein fragments: one able to bind to DNA and the other responsible for its hydrolysis [24–26]. Chimeric restriction enzymes were obtained by conjugating a zinc finger DNA-binding domain and the non-specific DNA-cleavage domain from the natural restriction enzyme Fok I. Zinc fingers are rather flexible in DNA recognition, thus by tuning their selectivity it was possible to cleave DNA in a sequence-specific manner (Fig. 6.3, see p. 230). This "proof of principle" opens the way for the generation of "artificial" nucleases that may cut DNA near a predetermined site.

Because of the high activity of metal ion complexes in catalyzing the hydrolytic cleavage of phosphates (see Section 6.2) Franklin and her group [27, 28] have designed chimeric 33- and 34-residue peptides which comprise helix 2 and 3 of engrailed homeodomain (the HTH region), and the 12-residue calcium-binding loop of an EF-hand from calmodulin. This protein binds quite strongly to Ca^{2+} which can be exchanged for a Ln(III) leading to equally robust metal complexes. Ln(III) ions are among the most powerful Lewis acid catalysts for phosphate cleavage [29]. Franklin's approach was to endeavor to obtain Ln(III)-binding peptides which were able to act as catalysts for the cleavage of bis-*p*-nitrophenylphosphate (BPNPP, a model for a DNA phosphate bond) and for the cleavage of DNA. In fact this turned out to be the case and she has shown that HTH/EF-hand chimeras bind lanthanides (specifically Eu^{3+}), have metal-dependent structures in solution, and cleave BPNPP and DNA.

All the above systems were based on fragments of native proteins or on peptide sequences prepared by copying the sequence of a known natural peptide. The design of totally synthetic sequences which are nevertheless active as catalysts is a challenging problem. Although different laboratories have shown that even very short di- or tripeptides form metal complexes which are hydrolytically active in the cleavage of model substrates [30–32], we will focus our attention on more complex sequences where the activity is not just the result of the mere formation of a metal complex. Barton and her group [33] reported the activity of zinc-binding peptides tethered to rhodium intercalators. In these systems the intercalator provides DNA binding affinity while the metal-binding peptide contributes the reactivity. The strategy appears to be rather general as Zn(II)-promoted DNA cleavage was observed for two widely different tethered metallopeptides. In one of these two metallopeptides, two imidazole units of histidines are facing each other in a α-helical conformation (Fig. 6.4, see p. 230), and this structural feature is directly involved in the binding of the metal ion. The other metallopeptide was modeled on the active site of the *Bam*HI endonuclease. Specifically, a short β hairpin was excerpted from *Bam*HI restriction endonuclease and the sequence was such that three carboxylate groups (one from an aspartic and two from glutamic acids) were clustered together thus facilitating metal ion binding. The native protein is known to bind to DNA as a dimer and cleaves at 5′-GGATCC-3′ palindromic sites to give four-base 5′ staggered ends. Barton et al. have also studied other redox-active metals such as copper, to effect oxidative cleavage. Their results indicate that one of the critical issues in the design of an artificial nuclease is the selection of the intercalator. What they have observed is that not all metallointercalators studied orient a metallopeptide for the initiation of DNA hydrolysis, and it may be necessary to vary linkers and ancillary ligands to optimize activity. The work carried out by Barton and others has shown that extreme caution should be taken in evaluating the mechanism of action by a metallocatalyst especially when the metal ion may also be involved in redox chemistry. In fact quite often what is assumed to be a hydrolytic mechanism eventually turns out to be oxidative. This is the case with metals such as copper, cobalt and, in selected cases, nickel as well.

Although the peptides described by Barton are highly active, it is difficult to control their conformation. In the case of the helical sequence, this structure amounted to no more than 30% of the total while in the case of the Rh-Bam conjugate the CD spectra did not show any characteristic features of helices or β hairpins. Certainly being able to control the conformation of the sequence in a more rigorous manner would make it easier to place the active groups in the putative catalytic site. With this in mind we have designed [34] a very simple sequence comprising two copies of a synthetic metal-binding amino acid (ATANP [6]) and five copies of Aib (a-amino isobutyric acid). Aib is the prototype of the class of C^a-tetrasubstituted amino acids that are known to impart helical conformation to even very short sequences [35] because of the conformational constraints imposed on the main chain by the geminal substituents at the alpha carbon. In the case of the heptapeptide shown in Fig. 6.5, the conformation is that of a 3_{10}-helix. This conformation differs from the a-helix by having a longer pitch (6.2 Å instead of 5.5 Å) and requiring only three amino acids rather than 3.6 in each turn of the helix. In this way the lateral arms of two amino acids placed in relative positions i and $i+3$ in the sequence face each other. This is exactly what happens to the aza-crowns of the two ATANP so that when they bind two metal ions (specifically, Zn^{2+}) these are placed in the correct position for cooperating in the cleavage of a phosphate bond which is a similar situation to that in hydrolytic enzymes. Indeed this was the case for the di-zinc complex of this peptide which turned out to be a very good catalyst for the hydrolysis of plasmid DNA with clear evidence of cooperativity between the two metal centers. The reactivity profile as a function of pH showed a maximum close to pH 7.3 supporting a mechanism in which one metal ion coordinates a phosphate anion while the other coordinates a water molecule. The conjugate base of the latter is the actual nucleophile. Thus the increase of activity up to pH 7.3 accounts for the deprotonation of this water molecule while the decrease in activity above this pH is related to the reduced ability of a phosphate to compete with a hydroxide for coordination to the metal center. This is a common feature of many catalysts with similar mechanisms.

Finally, it would be a serious omission if we failed to mention the totally different approach to the discovery of a metal ion-based peptide catalyst reported by Berkessel and Herault [36] in which they used the screening of a combinatorial library rather than rational design. The basic steps for the selection of active systems are as follows: (a) split and pool synthesis of the ligand library containing 625 solid-phase-bound undecapeptides; (b) complexation of the ligand library with Lewis acid transition metals (Cu^{2+}, Zn^{2+}, Fe^{3+}, Co^{3+}, Eu^{3+}, Ce^{4+}, Zr^{4+}); (c) screening of the library with chromogenic test substrates. This latter aspect was performed on beads, a challenging task that was accomplished by using a substrate that yielded a product which was easily oxidized by exposure to air with the formation of a colored, insoluble dye that adhered to the pellet holding the active cat-

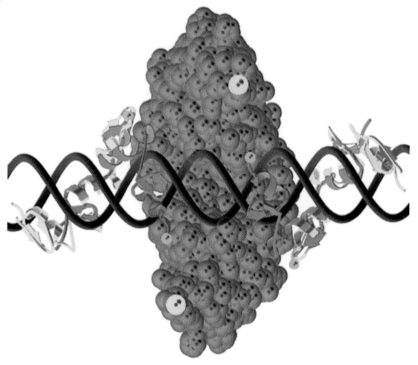

Fig. 6.3 Molecular model of the domains of the chimeric nuclease (constituted by an hybrid between a non-specific DNA cleavage domain and a zinc finger recognition domain) and DNA. The cleavage domain sits behind the DNA duplex while the zinc finger domain is shown in ribbon representation and winds through the major groove. (Reprinted from [26a] with permission from the American Society for Microbiology)

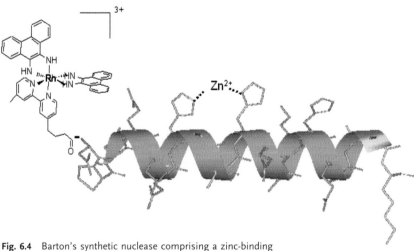

Fig. 6.4 Barton's synthetic nuclease comprising a zinc-binding peptide tethered to a rhodium intercalator [33]

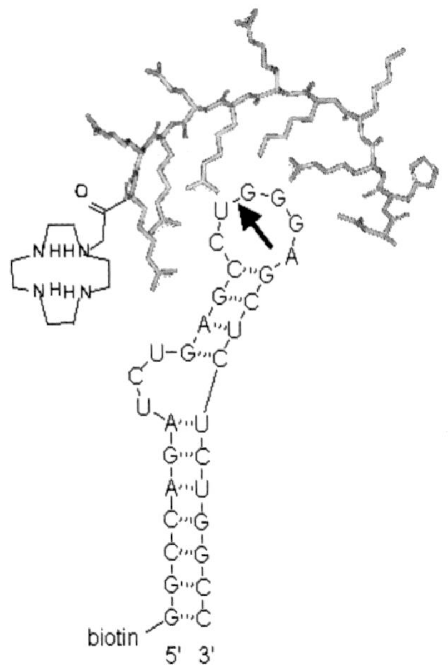

Fig. 6.7 Complex between 9mer peptide–cyclen conjugate and 31mer RNA as a model of TAR-RNA in HIV-1 infection as reported by Michaelis and Kalesse [48]

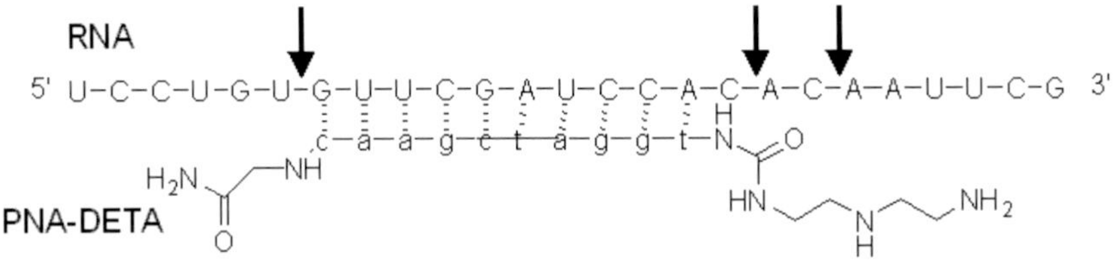

Fig. 6.8 PNA–diethylenetriamine conjugate bound to a 25mer RNA substrate [52]

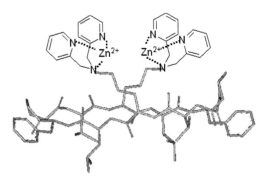

Fig. 6.9 Dinuclear Zn(II) complex of cyclic decapeptide as reported by Kawai et al. [53] modeled after the ionophoric cyclic peptide gramicidin S and active in the cleavage of HPNP

pertinent to this chapter, is that by Gunnlaugsson et al. [51]. The system they have described is again based on cyclen whose four nitrogens have been function-alized with glycine and uses a Ln(III) as a required cofactor. Since the reactivity was determined only with the model substrate HPNP, we do not know whether or not the system would operate preferentially on the basis of a metal-catalyzed mechanism with natural RNA.

The catalytic activity exerted by polyamines in the cleavage of RNA was also exploited by van Boom et al. [52]. They prepared a PNA–diethylenetriamine conju-gate (PNA-DETA) and demonstrated that it cleaved RNA at micromolar concentra-tions under physiological conditions and was very specific in its action (Fig. 6.8, see p. 235). The sequence of the PNA 10mer enabled it to bind to a 25mer RNA substrate characterized by a central sequence complementary to that of the puta-tive catalyst. Additional features of the substrate were the presence of two poten-tial neighboring C-A scission sites. Degradation studies were carried out at 40 °C and pH 7 and the analysis of the hydrolysis products revealed the presence of two major 5′-end labeled RNA fragments. Both fragments resulted from hydrolysis of the 3′-side of C17 and C19, respectively. The full cleavage of the substrate oc-curred in 24 h and, after 4 h almost 30% was already cleaved. The polyamine (polyammonium at the experimental pH) is required for catalysis as the same PNA sequence devoid of the DETA subunit is completely inactive. This work ele-gantly complemented the study by Kalesse mentioned above, where RNA recogni-tion was achieved with a small, natural peptide. Thus we have a short PNA that binds to the target substrate.

A recurrent feature of many of the catalysts described above is the importance of controlling the conformation. To tackle this problem Kawai et al. [53] have synthesized a cyclic decapeptide modeled on the ionophoric cyclic peptide gramici-din S (Fig. 6.9, see p. 235). The natural system possesses a stable antiparallel β-sheet conformation with two type II turns at the D-Phe-Pro sequences. Notably, the amino groups of the two Orn residues are located on one side of the β-sheet, and are suitable for the introduction of different functional groups. In fact the two amines could easily be functionalized by attaching to each of them two methylpyridine units to produce two metal ion-binding units which face each other. The activity of the dinuclear Zn^{2+} complex of this derivative was found to be extremely high with the rate of HPNP cleavage being very close to 4 orders of magnitude greater than the uncatalyzed reaction. The suggested mechanism of this two-metal ion-catalyzed reaction is very similar to that which we have sug-gested for the dimetallic peptide in Fig. 6.5 (see above). It appears rather obvious that in this particular example the conformation is controlled by the sequence of the peptide and the fact that it is cyclic, so that the number of accessible confor-mations is greatly diminished.

A number of less elaborate but, nevertheless, quite interesting systems have also been reported where just simple amino acids play a relevant role. Among them is an intriguing report by Komiyama's group [54] in which they describe RNA hydrolysis achieved as a result of the cooperation between carboxylate and ammonium ions. They prepared conjugates of glycine (or iminodiacetate) with anthraquinone and tested them in the hydrolytic cleavage of tRNA. The results indicate cooperativity between the carboxylate ion and ammonium ion on the basis of the following evidence. (a) Neither the ethyl ester nor the N-acetyl derivative of the glycine moiety are active in RNA hydrolysis; (b) a derivative devoid of the carboxylate but with only an amine showed no activity. The beauty of the system is mostly in its simplicity as the RNA binding unit (the anthraquinone) and the glycine unit, are the only constituents. It also shows that often the preparation of an effective catalyst does not require a complex design.

Along the same lines is a recent report by Orgel and Gao [55] who have shown that only a suspension of amino acids is required to hydrolyze tRNA. Their work is based on the observation that a variety of oligonucleotides and nucleic acids are strongly adsorbed from aqueous solutions onto the surfaces of crystals of many modestly soluble organic compounds. Thus the hydrolysis of tRNA is greatly accelerated by suspensions of aspartic and β-glutamic acids but not by suspensions of α-glutamic acid, asparagine or glutamine. They suggested that this specificity is the result of a detailed match between the surface structure and the RNA structure. Clear support for this suggestion is provided by the difference in behavior of D- and L-Asp. Since it is only the chirality of the solid which changes, it can therefore only be the orientation of the polymer with respect to the surface functional groups which is responsible for the differences. Indeed the surfaces of organic crystals comprise a very large ensemble of arrays of ordered organic functional groups as seen in a protein with a well-defined conformation. The authors concluded their work by asking a very intriguing question: "How often does a random surface approach a selected enzyme in binding affinity and catalytic power?" This is a serious matter for further thinking and could have implications for prebiotic chemistry and eventually, for the origin of life.

As the last example of the use of simple amino acids, we mention the report by Roth and Breaker [56] who have shown that histidine may act as a cofactor for a catalytic polynucleotide. Previously Arnold et al. [57] had shown that incorporation of a histidine unit as a linker in an antisense oligonucleotide promoted site-specific cleavage of RNA. The system of Breaker and Roth does not rely on metal ions for catalysis and is the result of *in vitro* selection [58]. This is an important finding because natural rybozimes require Mg^{2+} or other metal ion cofactors to induce formation of the correct structure (allosteric regulation) or to participate directly in the catalytic process [59]. The pool of DNA molecules used to initiate the selective-amplification process included a domain of 40 random-sequence nucleotides that was flanked on each side by regions of base complementation. Parallel selection was conducted by using reaction solutions buffered with histidine. The specificity of histidine as the cofactor was tested by using other amino acids and dipeptides. The mechanism of action of the most active mutants is quite similar to that of natural enzymes as they cleave RNA phosphoesters via a cyclizing mechanism to produce free 5'-hydroxyl and 2',3'-cyclic phosphate groups and the catalysis is general base.

6.5
Conclusions

Pseudo-peptides are involved in different ways in the catalysis of the cleavage of phosphate bonds. They may take part directly in the catalytic process or may be used as vehicles for the delivery of the catalytic entity to the appropriate target. The systems we have described represent just a portion of the considerable amount of work that has been published in this field. For instance the work by Dervan's group on pyrrole-imidazole polyamides as powerful and selective systems for binding to the minor groove of DNA [60, 61], although extremely relevant, has not been covered. In a broad sense their systems may be included in the class of pseudo-peptides. Their outstanding contribution to the field has been the development of hairpin molecules that uniquely recognize each of the four Watson–Crick base pairs while less emphasis has been put on the cleavage aspect.

In conclusion we have shown that peptide sequences may be used as ligands for metal ions either for oxidative or hydrolytic cleavage, as well as for the selective recognition of polynucleotides and the delivery of other cleavage agents to the target. In contrast, other approaches are based on the use of natural proteins or smaller fragments of them, taking advantage of a very efficient and specific interaction with DNA and/or RNA. In these cases the catalyst is made by conjugating these molecules with less specific phosphatases or with simple, synthetic cleavage catalysts. A few examples of attempts to obtain fully synthetic mini-enzymes by taking advantage of the peptide architecture for designing putative catalytic sites, mostly for hydrolytic, metal-catalyzed cleavage, have also been reported. The number of reports that continue to appear indicates that the interest of scientists in this field is still very high. Indeed the challenge of obtaining powerful nucleases, synthetic restriction enzymes or, in a general sense, synthetic molecules able to manipulate genes, is still open and success might be achieved in the not too distant future.

6.6
References

1 N. H. Williams, B. Takasaki, M. Wall, J. Chin, *Acc. Chem. Res.* **1999**, *32*, 485.

2 (a) A. Blask, T. C. Bruice, *Acc. Chem. Res.* **1999**, *32*, 475; (b) R. Wolfenden, C. Ridgway, G. Young, *J. Am. Chem. Soc.* **1998**, *120*, 833; (c) N. H. Williams, P. Wyman, *Chem. Commun.* **2001**, 1268.

3 H. At-Haddou, J. Sumaoka, S. L. Wiskur, J. F. Folmer-Andersen, E. V. Anslyn, *Angew. Chem. Int. Ed.* **2002**, *41*, 4013.

4 (a) D. E. Wilcox, *Chem. Rev.* **1996**, *96*, 2435; (b) N. Sträter, W. N. Lipscomb, T. Klabunde, B. Krebs, *Angew. Chem. Int. Ed. Engl.* **1996**, *35*, 2024.

5 (a) A. Torrado, B. Imperiali, *J. Org. Chem.* **1996**, *61*, 8940; (b) B. Imperiali, T. J. Prins, S. L. Fisher, *J. Org. Chem.* **1993**, *58*, 1613; (c) B. Imperiali, S. Fisher, *J. Am. Chem. Soc.* **1991**, *113*, 8527; (d) R. P. Cheng, S. Fisher, B. Imperiali, *J. Am. Chem. Soc.* **1996**, *118*, 11349; (e) H. Yamamoto, T. Nishina, H. Nishio, N. Yumoto, Y. Tatsu, T. Taguchi, S. Yoshikawa, *Peptide Chem.* **1994**, *41*, 1993; (f) N. Voyer, *J. Am. Chem. Soc.* **1991**, *113*, 1818; (g) J.-P. Mazaleyrat, A. Gaucher, Y. Goubard, J. Šavrada, M. Wekselman, *Tetrahed. Lett.* **1997**, *38*, 2091.

6 P. Rossi, F. Felluga, P. Scrimin, *Tetrahed. Lett.* **1998**, *39*, 7159.

7 H. Ishida, Y. Inoue, *Rev. Heteroat. Chem.* **1999**, *19*, 79.

8 C.B. Chen, L. Milne, R. Landgraf, D.M. Perrin, D.S. Sigman, *Chembiochem.* **2001**, *2*, 735.

9 J.S. Lazo. Bleomycin. In *Cancer Chemotherapy and Biological Response Modifiers Vol 18*, H.M. Pinedo, D.L. Longo, B.A. Chabner (Eds). Elsevier: New York, **1999**, pp 39–45.

10 A.T. Abraham, J.-J. Lin, D.L. Newton, S. Rybak, S.M. Hecht, *Chem. Biol.* **2003**, *10*, 45.

11 Latest examples: (a) S.T. Hoehn, H.-D. Junker, R.C. Bunt, C.J. Turner, J.A. Stubbe, *Biochemistry* **2001**, *40*, 5894; (b) C.J. Thomas, M.M. McCormick, C. Vialas, Z-F. Yao, C.J. Leitheiser, M.J. Rishel, X. Wu, S.M. Hecht, *J. Am. Chem. Soc.* **2002**, *124*, 3875.

12 E.C. Long, *Acc. Chem. Res.* **1999**, *32*, 827.

13 X. Huang, M.E. Pieczko, E.C. Long, *Biochemistry* **1999**, *38*, 2160.

14 (a) J.G. Muller, L.A. Kayser, S.J. Paikoff, V. Duarte, N. Tang, R.J. Perez, S.E. Rokita, C.J. Burrows, *Coord. Chem. Rev.* **1999**, *185–186*, 761; (b) A.J. Stemmler, C.J. Burrows, *J. Am. Chem. Soc.* **1999**, *121*, 6956.

15 D.P. Mack, B.L. Iverson, P.B. Dervan, *J. Am. Chem. Soc.* **1988**, *110*, 7572.

16 T.W. Bruice, J.G. Wise, D.S.E. Rosser, D.S. Sigman, *J. Am. Chem. Soc.* **1991**, *113*, 5446.

17 M. Nagaoka, M. Hagihara, J. Kuwahara, Y. Sugiura, *J. Am. Chem. Soc.* **1994**, *116*, 4085.

18 K.D. Copeland, A.M.K. Lueras, E.D.A. Stemp, J.K. Barton, *Biochemistry* **2002**, *41*, 12785.

19 K.E. Erkkila, D.T. Odom, J.K. Barton, *Chem. Rev.* **1999**, *99*, 2777.

20 D. Qi, C.-M. Tann, D. Haring, M.D. Distefano, *Chem. Rev.* **2001**, *101*, 3081.

21 D. Pei, P.G. Schultz, *J. Am. Chem. Soc.* **1991**, *113*, 9398.

22 (a) R.N. Zuckerman, P.G. Schultz, *Proc. Natl Acad. Sci. USA* **1989**, *86*, 1766 and references therein; (b) D. Pei, D.R. Corey, P.G. Schultz, *Proc. Natl Acad. Sci. USA* **1990**, *87*, 9858.

23 J.C. Truffert, U. Asseline, A. Brack, N.T. Thuong, *Tetrahedron* **1996**, *52*, 3005.

24 J. Smith, M. Bibikova, F.G. Whitby, A.R. Reddy, S. Chandrasegaran, D. Carroll, *Nucleic Acid Res.* **2000**, *17*, 3361.

25 Y. Kim, J. Cha, S. Chandrasegaran, *Proc. Natl Acad. USA* **1996**, *93*, 1156.

26 (a) M. Bibikova, D. Carroll, D.J. Segal, J.K. Trautman, J. Smith, Y.-G-Kim, S. Chandrasegaran, *Mol. Cell. Biol.* **2001**, *21*, 289; (b) M. Bibikova, M. Golic, K.G. Golic, D. Carroll, *Genetics* **2002**, *161*, 1169.

27 J.T Welch, M. Sirish, K.M. Lindstrom, S.J. Franklin, *Inorg. Chem.* **2001**, *40*, 1982.

28 J.T. Welch, W.R. Kearney, S.J. Franklin, *Proc. Natl Acad. Sci. USA* **2003**, *100*, 3725.

29 (a) S.J. Franklin, *Curr. Opin. Chem. Biol.* **2001**, *5*, 201; (b) M. Komiyama, *Chem. Commun.* **1999**, 1443.

30 (a) Y. Li, Y. Zhao, S. Hatfield, R. Wan, Q. Zhu, X. Li, M. McMills, Y. Ma, J. Li, K.L. Brown, C. He, F. Liu, X. Chen, *Bioorg. Med. Chem.* **2000**, *8*, 2675; (b) Y. Li, Y. Zhao, S. Hatfield, M. McMills, J. Li, X. Chen, *Bioorg. Med. Chem.* **2002**, *10*, 667.

31 C. Madhavaiah, S. Verma, *Bioconjug. Chem.* **2001**, *12*, 855.

32 R. Ren, W. Zheng, Z. Hua, P. Yang, *Inorg. Chem.* **2000**, *39*, 5454.

33 (a) M.P. Fitzsimons, J.K. Barton, *J. Am. Chem. Soc.* **1997**, *119*, 3379; (b) K.D. Copeland, M.P. Fitzimons, R.P. Houser, J.K. Barton, *Biochemistry* **2002**, *41*, 343.

34 C. Sissi, P. Rossi, F. Felluga, F. Formaggio, M. Palumbo, P. Tecilla, C. Toniolo, P. Scrimin, *J. Am. Chem. Soc.* **1999**, *121*, 6948.

35 (a) C. Toniolo, M. Crisma, F. Formaggio, G. Valle, G. Cavicchioni, G. Précigoux, A. Aubry, J. Kamphuis, *Biopolymers* **1993**, *33*, 1061; (b) M. Crisma, W. Bisson, F. Formaggio, Q.B. Broxterman, C. Toniolo, *Biopolymers* **2002**, *64*, 236.

36 A. Berkessel, D.A. Hèrault, *Angew. Chem. Int. Ed. Engl.* **1999**, *38*, 102.

37 V.N. Silnikov, V.V. Vlassov, *Russ. Chem. Rev.* **2001**, *70*, 491.

38 H. H. Thorp, *Chem. Biol.* **2000**, *7*, 33.

39 (a) B. Barbier, A. Brack, *J. Am. Chem. Soc.* **1988**, *110*, 6880; (b) B. Barbier, A. Brack, *J. Am. Chem. Soc.* **1992**, *114*, 3511.

40 B. Gutte, *J. Biol. Chem.* **1975**, *250*, 889.

41 B. Gutte, M. Daumigen, E. Wittshieber, *Nature* **1979**, *281*, 650.

42 W. Saenger, J. Riecke, D. Suck, *J. Mol. Biol.* **1975**, *93*, 529.

43 C.-H. Tung, Z. Wei, M.J. Leibowitz, S. Stein, *Proc. Natl Acad. USA* **1992**, 7114.

44 W.F. Lima, S.T. Crooke, *Proc. Natl. Acad. Sci. USA* **1999**, *96*, 10010.

45 A. Scarso, U. Scheffer, M. Göbel, Q.B. Broxterman, B. Kaptein, F. Formaggio, C. Toniolo, P. Scrimin, *Proc. Natl Acad. Sci. USA* **2002**, *99*, 5144.

46 S. Shinkai, M. Ikeda, A. Sugasaki, M. Takeuchi, *Acc. Chem. Res.* **2001**, *34*, 494.

47 (a) P. Rossi, F. Felluga, P. Tecilla, F. Formaggio, M. Crisma, C. Toniolo, P. Scrimin, *J. Am. Chem. Soc.* **1999**, *121*, 6948; (b) P. Rossi, F. Felluga, P. Tecilla, F. Formaggio, M. Crisma, C. Toniolo, P. Scrimin, *Biopolymers* **2000**, *55*, 496.

48 K. Michaelis, M. Kalesse, *Angew. Chem. Int. Ed. Engl.* **1999**, *38*, 2243.

49 Lysine rich peptides are known to bind to DNA, see for instance: R.C. Bergstrom, L.D. Mayfield, D.R. Corey, *Chem. Biol.* **2001**, *8*, 199.

50 K. Michaelis, M. Kalesse, *ChemBiochem.* **2001**, *1*, 79.

51 T. Gunnlaugsson, R.J.H. Davies, M. Nieuwenhuyzen, C.S. Stevenson, R. Viguier, S. Mulready, *Chem. Commun.* **2002**, 2136.

52 J.C. Verheijen, B. Deiman, E.Y. Yeheskiely, G.A. van der Marel, J.H. Van Boom, *Angew. Chem. Int. Ed. Eng.* **2000**, *39*, 369.

53 K. Yamada, Y. Takahashi, H. Yamamura, K. Saito, M. Kawai, *Chem. Comm.* **2000**, 1315.

54 M. Endo, K. Hirata, T. Ihara, S. Sueda, M. Takagi, M. Komiyama, *J. Am. Chem. Soc.* **1996**, *118*, 5478.

55 K. Gao, L.E. Orgel, *Helv. Chim. Acta* **2001**, *84*, 1347.

56 A.R. Roth, R.R. Breaker, *Proc. Natl Acad. USA* **1998**, *95*, 6027.

57 M.A. Reynolds, T.A. Beck, P.B. Say, D.A. Schwartz, B.P. Dwyer, W.J. Daily, M.M. Vaghefi, M.D. Metzler, R.E. Klem, L.J.A. Jr Arnold, *Nucleic Acid Res.* **1996**, *24*, 760.

58 R.R. Breaker, *Chem. Rev.* **1997**, *97*, 371.

59 A.M. Pyle, *Science* **1993**, *261*, 709.

60 See for instance: (a) Y.D. Wang, J. Dziegielewski, A.Y. Chang, P.B. Dervan, T.A. Beerman, *J. Biol. Chem.* **2002**, *277*, 42431; (b) A.Y. Chang, P.B. Dervan, *J. Am. Chem. Soc.* **2000**, *122*, 4856; (c) N.R. Wurtz, P.B. Dervan, *Chem. Biol.* **2000**, *7*, 153; (d) P.B. Dervan, R.W. Burli, *Curr. Opin. Chem. Biol.* **1999**, *3*, 688.

61 Recent related work by another group: T. Bando, A. Narita, I. Saito, H. Sugiyama, *Chem. Eur.J.* **2002**, *8*, 4781.

Subject Index